Natural Language Processing and Computational Linguistics 2

Series Editor
Patrick Paroubek

Natural Language Processing and Computational Linguistics 2

Semantics, Discourse and Applications

Mohamed Zakaria Kurdi

iSTE

WILEY

First published 2017 in Great Britain and the United States by ISTE Ltd and John Wiley & Sons, Inc.

ISTE Ltd
27-37 St George's Road
London SW19 4EU
UK

www.iste.co.uk

John Wiley & Sons, Inc.
111 River Street
Hoboken, NJ 07030
USA

www.wiley.com

Library of Congress Control Number: 2017953290

British Library Cataloguing-in-Publication Data
A CIP record for this book is available from the British Library
ISBN 978-1-84821-921-2

MIX
Paper from responsible sources
FSC® C013604

Contents

Introduction . . . ix

Chapter 1. The Sphere of Lexicons and Knowledge . . . 1

1.1. Lexical semantics . . . 1
1.1.1. Extension of lexical meaning . . . 1
1.1.2. Paradigmatic relations of meaning . . . 6
1.1.3. Theories of lexical meaning . . . 16
1.2. Lexical databases . . . 23
1.2.1. Standards for encoding and exchanging data . . . 25
1.2.2. Standard character encoding . . . 25
1.2.3. Content standards . . . 32
1.2.4. Writing systems . . . 40
1.2.5. A few lexical databases . . . 45
1.3. Knowledge representation and ontologies . . . 49
1.3.1. Knowledge representation . . . 49
1.3.2. Ontologies . . . 63

Chapter 2. The Sphere of Semantics . . . 75

2.1. Combinatorial semantics . . . 75
2.1.1. Interpretive semantics . . . 75
2.1.2. Generative semantics . . . 80
2.1.3. Case grammar . . . 82
2.1.4. Rastier's interpretive semantics . . . 84
2.1.5. Meaning–text theory . . . 92

2.2. Formal semantics 95
2.2.1. Propositional logic 95
2.2.2. First-order logic 106
2.2.3. Lambda calculus 113
2.2.4. Other types of logic 121

Chapter 3. The Sphere of Discourse and Text 123

3.1. Discourse analysis and pragmatics 123
3.1.1. Fundamental concepts 123
3.1.2. Utterance production 125
3.1.3. Context, cotext and intertextuality 128
3.1.4. Information structure in discourse 130
3.1.5. Coherence 137
3.1.6. Cohesion 138
3.1.7. Ellipses 142
3.1.8. Textual sequences 143
3.1.9. Speech acts 144
3.2. Computational approaches to discourse 146
3.2.1. Linear segmentation of discourse 146
3.2.2. Rhetorical structure theory and automatic discourse analysis 148
3.2.3. Discourse interpretation: DRT 154
3.2.4. Processing anaphora 159

Chapter 4. The Sphere of Applications 169

4.1. Software engineering for NLP software 169
4.1.1. Lifecycle of an NLP software 169
4.1.2. Software architecture for NLP 170
4.1.3. Serial architectures 171
4.1.4. Data-centered architectures 173
4.1.5. Object-oriented architectures 177
4.1.6. Multi-agent architectures 178
4.1.7. Syntactic–semantic cooperation: from cognitive models to software architecture 180
4.1.8. Programming languages for NLP 184
4.1.9. Evaluation of NLP systems 186
4.2. Machine translation (MT) 191
4.2.1. Why is translation difficult? 192
4.2.2. History of MT systems 194
4.2.3. Typology of MT systems 196

4.2.4. The use of MT 198
4.2.5. MT techniques 199
4.2.6. Example of a translation system: Verbmobil 208
4.3. Information retrieval (IR) 211
4.3.1. IR and related domains 211
4.3.2. Lexical information and IR 213
4.3.3. Information retrieval approaches 219
4.4. Big Data (BD) and information extraction 234
4.4.1. Structured, semi-structured and unstructured data 234
4.4.2. Architectures of BD processing systems 235
4.4.3. Role of NLP in BD processing 237
4.4.4. Information extraction 238

Conclusion 259

Bibliography 263

Index 301

Introduction

Language is a central tool in our social and professional lives. It is a means to convey ideas, information, opinions and emotions, as well as to persuade, request information, give orders, etc. The interest in language from a computer science point of view began with the start of computer science studies themselves, notably in the context of work in the area of artificial intelligence. The Turing test, one of the first tests developed to determine whether a machine is intelligent or not, stipulates that to be considered as intelligent, the machine must have conversational capacities comparable to those of a human [TUR 50]. This means that an intelligent machine must have the capacity for comprehension and generation, in the broad sense of the terms, hence the interest in natural language processing (NLP) at the dawn of the computer age. Historically, computer processing of languages was very quickly directed toward applied domains such as machine translation (MT) in the context of the Cold War. Thus, the first MT system was created as part of a shared project between Georgetown University and IBM in the United States [DOS 55, HUT 04]. These applied works were not as successful as intended and the researchers quickly became aware that a deep understanding of the linguistic system was a prerequisite for any successful application.

The internet wave between the mid-1990s and the start of the 2000s was a very significant driving force for NLP and related domains, notably information retrieval, which grew from a marginal domain limited to information retrieval in the context of a large company to information retrieval on the scale of the Internet, whose content is constantly growing. This development in terms of the availability of data also favored a discipline that was already in its infancy: Data Science. Located at the intersection of statistics, computer science and mathematics, Data Science focuses on the analysis, visualization and processing of digital data in all forms: images, text and speech. The role of NLP within Data Science is obvious, given that the majority of the information processed is contained in written documents or speech recordings. It is therefore possible to distinguish two different but complementary research approaches in the domain of NLP. On the one hand, there are works that aim to solve the fundamental problem of language processing and that are consequently concerned with the cognitive and linguistics aspects of this problem. On the other hand, several works are dedicated to optimizing and adapting existing NLP techniques for various applied domains such as the medical or banking sectors.

The objective of this book is to provide a comprehensive review of classic and modern works in the domains of lexical databases and the representation of knowledge for NLP, semantics, discourse analysis, and NLP applications such as machine translation and information retrieval. This book also aims to be profoundly interdisciplinary by giving equal consideration to linguistic and cognitive models, algorithms and computer applications as much as possible because we are starting from the premise, which has been proven in NLP and elsewhere time and time again, that the best results are the product of a good theory paired with a well-designed empirical approach.

In addition to the Introduction, this book has four chapters. The first chapter concerns the lexicon and the representation of knowledge. After an introduction to the principles of lexical semantics and theories of lexical meaning, this chapter covers lexical databases, the main procedures for representing knowledge and ontologies. The second chapter is dedicated to semantics. First, the main approaches in combinatorial semantics such as interpretive semantics, generative semantics, case grammar, etc. will be presented. The next section is dedicated to the logical approaches to formal semantics used in the domain of NLP. The third chapter focuses on discourse. It covers the fundamental concepts in discourse analysis such as utterance production, thematic progression, structuring information in discourse, coherence and cohesion. This chapter also presents different approaches to discourse processing such as linear segmentation, discourse analysis and interpretation, and anaphora resolution. The fourth and final chapter is dedicated to NLP applications. First, the fundamental aspects of NLP systems such as software architecture and evaluation approaches are presented. Then, some particularly important applications in the domain of NLP, such as machine translation, information retrieval and information extraction, are reviewed.

1

The Sphere of Lexicons and Knowledge

1.1. Lexical semantics

Located at the intersection of semantics and lexicology, lexical semantics is a branch of semantics that focuses on the meaning of words and their variations. Many factors are taken into consideration in these studies:

– The variations and extensions of meaning depending on the usage context. The context can be linguistic (e.g. surrounding words), which is why some experts call it "cotext" instead. The context can also be related to the use or register of the language. In this case, it can indicate the socio-cultural category of the interlocutors, for example, formal, informal or vulgar.

– The semantic relationship that the word has with other words: synonyms, antonyms, similar meaning, etc. The grammatical and morphological nature of these words and their effects on these relationships are also of interest.

– The meaning of words can be considered to be a fairly complex structure of semantic features that each plays a different role.

This section will focus on the forms of extension of lexical meaning, the paradigmatic relations between words and the main theories concerning lexical meaning.

1.1.1. *Extension of lexical meaning*

Language users are aware that the lexical units *to fight, to rack one's brain* and *crazy* are not used literally in sentences [1.1]:

– The minister fought hard to pass the new law. [1.1]

– Mary racked her brain trying to find the street where John lived.

– John drives too fast, he's crazy!

Far from the simplicity of the everyday use of these descriptive uses, the figurative use of lexical items occurs in different forms that will be discussed in the following sections.

1.1.1.1. *Denotation and connotation*

From the perspective of the philosophy of language, *denotation* designates the set of objects to which a word refers. From a linguistic point of view, denotation is a stable and objective element because it is shared, in principle, by the entire linguistic community. This means that denotation is the guarantee of the conceptual content of the lexicon of a given language.

Depending on the context, *connotation* is defined as the set of secondary significations that are connected to a linguistic sign and that are related to the emotional content of the vocabulary. For example, the color red denotes the visual waves of certain physical properties. Depending on the context, this color has several different connotations. Here are a few linguistic contexts where the word *red* can be used with their connotations (see Table 1.1).

Context	Connotation
Glowing red coals.	very hot
The streets ran red after the conflict.	blood-stained
John offered Mary a red rose.	love
Red light.	interdiction
Red card (Soccer).	expulsion
The Red Scare (politics).	Communism

Table 1.1. *Examples of the connotations of the color red*

In some cases, the difference between connotation and denotation pertains to the register of language. For example, the groups (dog, mutt, pooch), (woman, chick), (police officer, cop, pig) each refer to the same object but the words of each group have different connotations that can provide information about the socio-cultural origins of the interlocutor and/or the situation of communication.

The distinction between denotation and connotation is considered to be problematic by some linguists. Linguistic evolution means that external features or properties become ingrained over time. For example, the word *pestilence*, which refers to an illness, has evolved with time and now also refers to a disagreeable person, as in: Mary is a little pest.

1.1.1.2. *Metaphor*

Of Greek origin, the word *metaphor* literally means transfer. It consists of the semantic deviation of a lexical item's meaning. Traditionally, it is a means to express a concept or abstract object using a concrete lexical item with which it has an objective or subjective relationship. The absence of an element of comparison such as *like* is what distinguishes metaphor from simile. The sentence *she is as beautiful as a rose* is an example of a simile.

There needs to be only some kind of resemblance for the metaphor process to enter into play. These resemblances can concern a property: *to burn with love* (intense and passionate love); the form: *a rollercoaster life* (a life with ups and downs like a rollercoaster ride), *John reaches for the stars* (to set one's sights high or be very ambitious), *genealogical tree* (a set of relations whose diagram is similar to the shape of the branches of a tree); the degree: *to die of laughter* (death is an extreme state); the period: *the springtime of life* (youth), *the Arab Spring* (renewal); or personification: *the whale said to Sinbad, "You must go in this direction"* (the whale spoke like a person).

In some cases, there are objects that do not have a proper designation (non-lexicalized objects). They metaphorically borrow the names of other objects. This includes things like the wing of a plane, a windmill or a building, which all borrow the term of a bird's limb because of the resemblance in terms of form or function. This metaphor is called a catachresis.

From a cognitive perspective, there are two opposing schools of thought when it comes to the study of metaphors: the constructivist movement and the non-constructivist movement. According to the constructivist movement, the objective world is not directly accessible. It is constructed on the basis of restricting influences on both language and human knowledge. In this case, metaphor can be seen as an instrument used to construct reality. According to the Conceptual Metaphor Theory of [LAK 80, LAK 87], the most extreme form of constructivism, metaphor is not a stylistic decorative effect at all Rather, it is an essential component of our cognitive system that allows us to concretely conceptualize an abstract idea. The basic idea of this theory is that a metaphor is a relationship of correspondence between two conceptual domains: the source domain and the destination domain. According to this theory, metaphor is not limited to a particular linguistic expression because the same metaphor can be expressed in several different ways. To illustrate this idea, Lakoff gives the example of the metaphor of the voyage of life, where life is the source domain and the voyage is the destination domain (see Table 1.2).

Life	Voyage
Birth	Start of the voyage
Death	End of the voyage
Reaching an objective	Arriving at the destination
Point of an important choice	Intersection
Difficulties	Obstacles
Encountering difficulties	Climbing
Colleagues, friends, partners, etc.	Co-travelers

Table 1.2. *The metaphor of life as a voyage*

The correspondences presented in Table 1.2 are the source of expressions like "It's the end of the road for John", and "Mary is progressing quickly but she still has not arrived at the point where she wants to be", etc. Note that in Lakoff's approach, two types of correspondences are possible: ontological correspondences that involve entities from different domains and epistemological correspondences that involve knowledge about entities.

As shown in [LAK 89], the correspondences are unidirectional even in the case of different metaphors that share the same domain. They give the example of humans as machines and machines as humans (see Table 1.3).

Humans as machines	Machines as humans
John is very good at math, he's a human calculator. Marcel is a harvesting machine.	I think my car doesn't like you, she doesn't want to start this morning. The machine did not like the new engine, it was too weak. My computer told me that the program wasn't working.

Table 1.3. *The metaphor of humans as machines and machines as humans*

Although these metaphors share the same domain, the features used in one direction are not the same as the features used in the other direction. For example, in the metaphor of humans as machines, the functional features associated with machines are efficiency, rapidity and precision, projected onto humans. On the other hand, different features like desire and the capacity for communication are projected onto machines.

Metaphors are far from being considered a marginal phenomenon by linguists. In fact, some believe that studying metaphorical language is fundamental for understanding the mechanisms of language evolution because many metaphors pass into ordinary use. Other models have also been proposed, including the theory of lexical facets [KLE 96, CRU 00, CRU 04].

1.1.1.3. *Metonymy*

Metonymy consists of designating an object or a concept by the name of another object or concept. There are different types of metonymy depending on the nature of the connections that relate the objects or concepts:

– The cause and its effect: the harvest can designate the product of the harvest as well as the process of harvesting.

– The container for the contents: *he drank the whole <u>bottle</u>, he ate the whole <u>box/plate</u>.*

– The location for the institution that serves there: *The Pentagon decided to send more soldiers into the field. Matignon decided to make the documents public* (Matignon is the castle where the residence and the office of the French Prime Minister is located in Paris).

Like metaphors, the context plays an important role in metonymy. In fact, sentences like *I have read Baudelaire* (meaning that I have read poems written by Baudelaire) can only be interpreted as metonymies because the verb *to read* requires a readable object (e.g. a book, newspaper, novel, poem). Since the object here is a poet, we imagine that there is a direct relationship with what we have read his poems.

1.1.1.4. *Synecdoche*

Synecdoche, a particular case of metonymy, consists of designating an object by the name of another object. The relationship between the two objects can be a varied form of inclusion. Here are a few examples:

– A part for the whole, as in: *the sails are close to port* (sails/ship), or *new hands join in the effort* (hands/person), or *the jaws of the sea* (jaws/shark).

– The whole for a part: *Italy won the European Cup* (Italy/Italian team).

– From the specific to the general: *Spring is the season of roses* (roses/all kinds of flowers).

As noted, unlike metonymy, the two objects involved in a synecdoche are always inseparable from one another.

1.1.2. *Paradigmatic relations of meaning*

Language is far from being a nomenclature of words. Words have varied relationships on different levels. In addition to syntagmatic relations of co-occurrence, which are fundamentally syntactical, words have essentially semantic paradigmatic relations. These relations can be linear, hierarchical, or within clusters.

1.1.2.1. *Semantic field and lexical field*

Used to designate the structure of a linguistic domain, the term *field*, while fundamental in lexicology, can refer to various concepts depending on the school of thought or linguists. Generally, following the German tradition of distinction between *sinnfeld* (field of meaning) and *wortfeld* (field of words), there is a distinction made between the lexical field and the semantic field [BAY 00]. A lexical field is defined as a set of words that pertain to the same domain or the same sector of activity. For example, the words *raid*, *anti-tank, armored vehicle*, *missile* and *machine gun* belong to the lexical field of war. In cases of polysemy, the same word belongs to several fields. For example, the word *operation* belongs to these three fields: mathematics, war and medicine [MIT 76][1]. The semantic field is defined as the area covered by the signification(s) of a word in a language at a given moment in its history [FUC 07]. In this regard, the semantic field is related to polysemy. Faced with this terminological confusion, two approaches from two linguistic currents proposed representing polysemes in terms of their shared meaning. The first approach, presented by Bernard Pottier and François Rastier, is part of the structural semantics movement and analyzes according to the hierarchy of semantic components: taxeme, domain, dimension (see section 2.11 on the interpretive semantics of Rastier). The second approach, presented by Jacqueline Picoche, falls under the context of Gustave Guillaume's psychomechanics and proposes lexical-semantic fields [PIC 77, PIC 86].

As underscored in [CRU 00], the relations between the terms in a field are hierarchical. They follow the diagram shown in Figure 1.1.

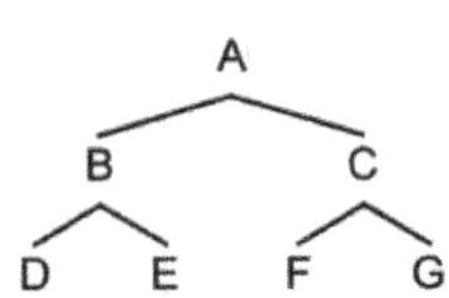

Figure 1.1. *General diagram of lexical hierarchies in a field [CRU 00]*

1 Note that what is called the lexical field here is called the semantic field in [MIT 76].

As can be seen in Figure 1.1, two types of relations emerge from lexical hierarchies: relations of dominance, like the relationships between *A* and (*B*, *C*) or *B* and (*D*, *E*) and relations of differentiation, such as the relationships between *B* and *C* or *F* and *G*. From a formal perspective, the trees are acyclic-directed graphs (there is no path with points of departure or arrival). In other words, if there is a link between two points *x* and *y*, then there is no link in the inverse direction[2]. Furthermore, each node has a single element that immediately dominates it, called the parent node, and potentially it has one or more child nodes itself.

In lexical hierarchies, the symbols *A*, *B*, ...*G* correspond to lexical items. Cruse distinguishes between two types of hierarchies: taxonomic, or classificatory, hierarchies and meronymic hierarchies.

1.1.2.2. *Taxonomic hierarchies*

These hierarchies reflect the categorization of objects in the real world by members of a given linguistic community. First, consider the example of the classification of animals presented in Figure 1.2.

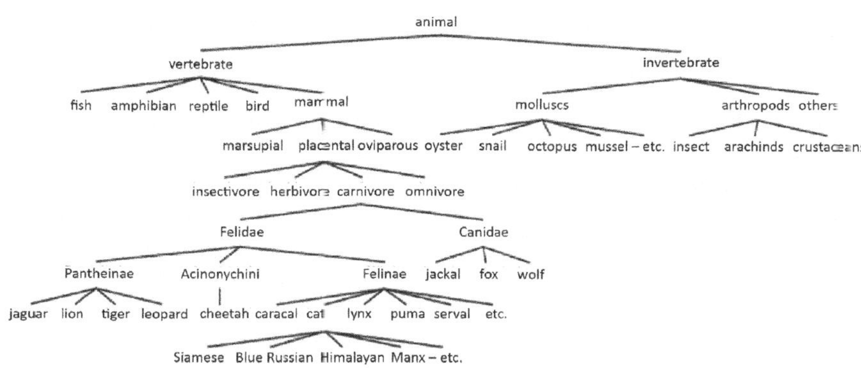

Figure 1.2. *Partial taxonomy of animals*

In a taxonomic hierarchy, the element at the higher classification level, or the parent, is called the hyperonym and the lower element, the child, is called the hyponym. Thus, *animal* is the hyperonym of *fish* and *Felidae* is the hyponym of *carnivore*. They mark a level of genericity or precision, as in the following exchange [1.2].

2 The link represents a unidirectional semantic relation (e.g. *type of*). There can be non-directed graphs with two opposite relations on each arch following the navigation direction (type of/class of).

Did you buy apples at the market? [1.2]

Yes, I bought a kilogram of Golden Delicious.

In exchange [1.2], *Golden Delicious*, hyponym of *apple*, is used here to give more specific information in response to the question. The inverse process could be used to hide some of the information.

The root is the basic element of the tree. It is distinguished by greater levels of genericity and abstraction than all other elements of the tree. Often, it is not a concrete object, but rather a set of features shared by all of the words in the group. In the example in Figure 1.2, the root element *animal* cannot be associated with a visual image or a given behavior. It is also important to note that the number of levels can vary considerably from one domain to another. According to [CRU 00], taxonomies related to daily life such as dishes and appliances tend to be flatter than taxonomies that pertain to the scientific domains. Some research has indicated that the depth of daily life taxonomies does not extend past six levels. Obviously, the depth of the tree depends on the genericity and the detail of the description. For example, a level above *animal* can be added if an expansion of the description is desired. Similarly, we can refine the analysis by adding levels that correspond to types of cats: with or without fur, domestic or wild, with or without tails, etc. There is a certain amount of subjectivity involved in taxonomic descriptions due to the level of knowledge of the domain as well as the objectives of the description.

Finally, it is also useful to mention that certain approaches, especially those of the structural current, prefer to expand the tree with distinctive features that make it possible to differentiate elements on the same level. For instance, the feature [+vertebral column] and [–vertebral column] could be placed on *vertebrate* and *invertebrate*, respectively. Similarly, the feature: [aquatic] and [cutaneous respiration] can be used to distinguish fish from amphibians.

1.1.2.3. *Meronymic hierarchies*

Meronymic and holonymic relations are the lexical equivalents of the relationship between an object and its components: the components and the composite. In other words, they are based on relations like *part of* or *composed of*. In a meronymic tree, the parent of an element is its holonym and the child of an element is its meronym.

Some modeling languages, like the Unified Modeling Language (UML), distinguish between two types of composition: a strong composition and a weak composition. Strong composition concerns elements that are indispensable to an entity, while weak composition pertains to accessories. For example, a car is not a car without wheels and an engine (strong composition) but many cars exist that do not have air conditioning or a radio (weak composition). This leads to another distinction between strong meronymy and weak meronymy. In the case of strong meronymy, the parts form an indissociable entity. Weak meronymy connects objects that can be totally independent but form an assorted set. For example, a suit must be made up of trousers and a jacket (strong composition). Sometimes, there is also a vest (weak composition). For varied and diverse reasons, the trousers can be worn independent of the jacket and vice versa. However, this kind of freedom is not observed concerning the wheel or the engine of a car, which cannot be used independently of the car, the entity they compose.

An interesting point to mention concerning the modeling of these relations is the number of entities involved in the composition relation, both on the side of the components and on the side of the composites, which are commonly called the multiplicity and the cardinality of the relation, respectively. Thus, it is worth mentioning that a human body is composed of a single heart and that any one particular heart only belongs to one body at a time, in a one-to-one relation. Similarly, the body has a one-to-two cardinal relationship with eyes, hands, feet, cheeks, etc. The cardinal relationship between a car and a wheel is one-to-many, because a car has several wheels (four or sometimes more).

Figure 1.3 presents a hierarchy of body parts with a breakdown of the components of the head.

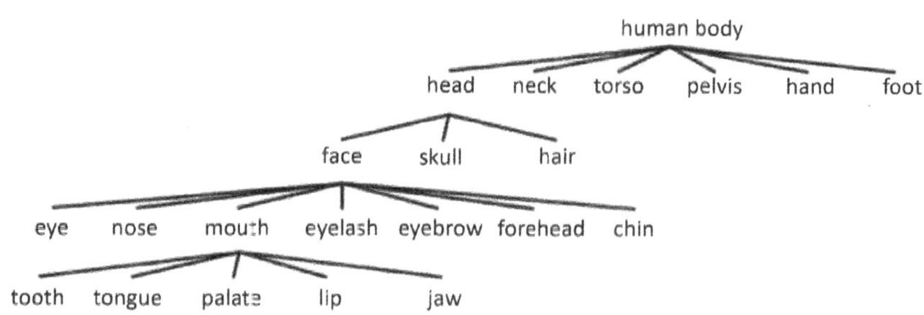

Figure 1.3. *Meronymic hierarchy of the human body*

Even more so than in the case of taxonomic hierarchies, there is no absolute rule in this kind of hierarchy to decide if an element is a part of an entity or not. For example, the neck could just as well be part of the head as part of the torso. The same goes for shoulders, which could be considered part of the arms or part of the torso.

1.1.2.4. *Homonymy and polysemy*

Homonymy is the relation that is established between two or more lexical items that have the same signifier but fundamentally different signifieds. For example, the verb *lie* in the sense: utter a false statement, and *lie* in the sense: to assume a horizontal position are homonyms because they share the same pronunciation and spelling even though there is no semantic link between them. There are also syntactic differences between the two verbs, as they require different prepositions to introduce their objects (*lie to someone* and *lie in*). In addition to rare cases of total homonymy, two forms of partial homonymy can be distinguished: homophony and homography.

Homophony is the relation that is established between two words that have different meanings but an identical pronunciation. Homophones can be in the same grammatical category, such as the nouns *air* and *heir* that are pronounced [er], or from different categories, like the verb *flew* and the nouns *flue or flu* that are pronounced [flü].

Homography is the relationship between two semantically, and some syntactically, different words that have an identical spelling. For example, *bass* [beɪs] as in: *Bass is the lowest part of the musical range*, and *bass* [bas] as in: *Bass is bony fish* are two homographs. Note that when homonymy extends beyond the word as in: *she cannot bear children*, this is often referred to by the term *ambiguity*.

Polysemy designates the property of a lexical item having multiple meanings. For example, the word *glass* has, among others, the following meanings: vitreous material, liquid or drink contained in a glass, a vessel to drink from and lenses. Once in a specific context, polysemy tends to disappear or at least to be reduced, as in these sentences [1.3]:

John wants to drink a glass of water.
John bought new glasses. [1.3]
Mary offered a crystal glass to her best friend.

It should also be noted that polysemy sometimes entails a change in the syntactic behavior of a word. To illustrate this difference, consider the different uses of the word *mouton* (sheep in French) presented in Table 1.4.

In the sentences presented in Table 1.4, the syntactic behavior of the word *mouton* varies according to the semantic changes.

Polysemy is part of a double opposition that is composed of monosemic units and homonymic units.

Jean a attrapé un petit mouton. Jean caught a little sheep.	Animal/DOC
Jean cuisine/mange du mouton. Jean cooks/eats sheep.	Meat/IOC
Jean possède une vieille veste de mouton. Jean has an old jacket made of sheep leather/skin.	leather/skin/noun complement

Table 1.4. *Examples of polysemy*

The first opposition is with monosemic lexical units that have a single meaning in all possible contexts. These are rare, and are often technical terms like: *hepatology*, *arteriosclerosis* and *hypertension*. Nouns used to designate species also have a tendency to be monosemic in their use outside of idiomatic expressions: rhinoceros, aralia, adalia, etc.

The second opposition, fundamental in lexicology and lexicography, is between homonymy and polysemy. The main question is: what criteria can be used to judge whether we are dealing with a polysemic lexical item or a pair of homonyms? The criterion used to determine that the original polysemy has been fractured, leaving in its place two different lexical entries that have a homonymic relationship, is the semantic distance perceived by speakers of the language. If, on the other hand, this link is no longer discernable, the words are considered to be homonyms. The issue with this criterion is that it leaves a great deal to subjectivity, which results in different treatments. In dictionaries, polysemy is presented in the form of different meanings for the same term, while distinct entries are reserved for homonyms. For example, the grapheme *bear* is presented under two different entries in the Merriam-Webster dictionary[3]: one for the noun (the animal) and one for the verb to move while holding up something. On the other hand, there is one entry for the word car with three different meanings (polysemy): *a vehicle moving on wheels*, *the passenger compartment of an elevator*, and *the part of an airship or balloon that carries the passengers and cargo*.

It should be noted that ambiguity can be seen as the other side of polysemy. In her book *Les ambiguïtés du français*, Catherine Fuchs considers that polysemy can also concern extra-lexical levels such as the sentence [FUC 96]. For example, in the sentence: *I saw the black magic performer*, the adjective *black* qualifies either the performer or magic.

3 https://www.merriam-webster.com/dictionary/bear

1.1.2.5. *Synonymy*

Synonymy connects lexical items of the same grammatical category that have the same meaning. More formally, in cases of synonymy, two signifiers from the same grammatical category are associated with the same signified. Synonymy exists in all languages around the world and corresponds to semantic overlap between lexical items. It is indispensable, particularly for style and quality. One way to determine synonymy is to use the method of interchangeability or substitution.

If two words are interchangeable in all possible contexts, then they are said to be a case of total or extreme synonymy. Rather rare, it especially concerns pairs of words that can be considered to be morphological variants, such as *ophtalmologue/ophtalmologiste* (opthalmologist in French), *she is sitting/she is seated.*

Partial synonymy occurs in cases of interchangeability limited to certain contexts. For instance, consider these pairs: *car/automobile*, *peril/danger*, *risk/danger*, *courage/bravery* and *distinguish/differentiate*. These pairs are interchangeable in certain (common contexts) and are not in others (distinctive contexts) (see Table 1.5). Polysemy constitutes the primary source of this limit of interchangeability, because often words have several meanings, each of which is realized in a precise context, where it is synonymous with one or several other words.

John drives (the car/gondola).	Common context
John drives his car to go to work (automobile). John drives his gondola to go to work. ?	Distinctive context
He wants to keep the company safe from all (dangers/peril/risks).	Common context
He lives in fear of the Yellow Peril. He lives in fear of the yellow danger/risk. ?	Distinctive context
It is not easy to (differentiate/distinguish) him from his brother.	Common context
The Goncourt Prize distinguished this extraordinary novel. The Goncourt Prize differentiated this extraordinary novel. ?	Distinctive context

Table 1.5. *Examples of partial synonyms*

The use of a lexical unit by a particular socio-cultural category can add a socio-semantic dimension to this unit, according to the terms of [MIT 76], which is then differentiated by other synonyms. For example, the following pairs are synonyms, but are distinguished by a different social usage (familiar or vulgar vs. formal): *guy/man*, *yucky/disgusting*, *boring/tiresome*. Geo-linguistic factors also play a role.

For example, in the east of France, the lexical unit *pair of* can be synonymous with the number *two* as in *a pair of birds* or *a pair of shoes* [BAY 00]. In everyday use, these words are not synonyms: *a pair of glasses* is not the same as *two glasses*.

Sometimes two lexical units can have the same denotation but two different connotations. For example, *an obese woman* and *a fat woman* both designate someone of the female sex who suffers from excessive weight but the phrases nevertheless have different connotations.

The use of a word in the context of a locution or a fixation is one of the reasons that limit its synonymic relations. For example, the word *risk* used in locutions such as *at risk* or *at risk of* makes it non-substitutable in these locutions with words that are otherwise its synonyms, like *danger* and *peril*. Similarly, the words *baked* and *warmed* that are synonyms in a context like *the sun baked the land/warmed the land* are no longer synonyms when *baked* is used in a fixed expression like *we baked the cake*.

Finally, synonymy does not necessarily imply a parallelism between two words. The French nouns *éloge* and *louange* (praise) are synonyms and so are the adjectives *louangeur* and *élogieux* (laudatory). As the morphological nature of these two words is different, the parallelism does not extend to the verbal form, given that the verb *élogier* does not exist in French to be the synonym of the verb *louanger*.

1.1.2.6. *Opposition*

The semantic nature of some lexical units logically involves a certain form of opposition. This makes opposition a phenomenon that is universally shared by all languages in the world. However, the definition of this relation is not simple. In fact, several forms of oppositions exist and, to determine a type, logical and linguistic criteria are often used.

The simplest form of opposition is called binary, polar or privative opposition. This concerns cases where there is no gradation possible between the opposed words. For example, between *dead* and *alive*, there is no intermediary state (*zombie* being a purely fictional state).

Oppositions that are gradable or scalar are distinguished by the existence of at least one intermediary or middle state. The opposition between the pairs *long/short*, *hot/cold* and *fast/slow* is gradable and allows for a theoretically infinite number of intermediary states.

To distinguish these two forms of opposition from other forms, a logical test can be applied that consists of a double negation according to these two rules:

$$A \rightarrow \neg B$$

$$\neg B \rightarrow A$$

A and *B* being two antonyms, → is a logical implication (→ is read as *if … then*) and ¬ is the negation symbol (¬B is read as not *B*). Applied to the pair *open/closed*, these rules provide the following inferences:

open → ¬ closed (if a door is open then it is not closed).

¬ closed → open (if a door is not closed then it is open).

Gradual oppositions do not validate the first rule. If a car is fast, that does not necessarily mean that it is not slow (it could be in any one of an innumerable intermediary states).

Pairs that designate equivalent concepts such as days of the week, months and metrological units can only validate the first rule:

April → ¬ July (if it is April, then it is not July)

¬ July → April *

From a linguistic point of view, we recognize adjectives through the possibility or impossibility of inserting them in front of an intensifier or using them as comparatives or superlatives. Adjectives such as *small*, *intelligent* and *fast* can often be used with an intensifier as in: *very fast*, *fairly intelligent* and *too small*. They can also be employed as comparatives and superlatives as in: the most intelligent, as fast as. Some linguists introduce degrees of nuance to the two large forms of opposition that we just discussed. For example, the oppositions *male/female*, *man/woman* or *interior/exterior* are traditionally considered to be a relation of **complementarity**. Some prefer to call the two extremes of a gradual opposition **antipodes** (peak/foot).

To visually represent the relations of opposition, [GRE 68] proposed the semiotic square. This is a process that makes it possible to logically analyze oppositions by considering logically possible classes that result from a binary opposition (see Figure 1.4).

7. Positive deixis		Complex term		8. Negative deixis
	Term A		Term B	
		9. Positive schema 10. Negative schema		
	Term not-B		Term not-A	
		Neutral term		

Figure 1.4. *General structure of a semiotic square*

Thus, the man/woman opposition can give rise to the classes: man, woman, man and woman (hermaphrodite or androgyne), neither man nor woman (person suffering from genital deformation). This can produce the semiotic square presented in Figure 1.5. [HEB 12].

macho		Masculine + feminine Hermaphrodite/androgynous		vamp
	Masculine (man)		Feminine (woman)	
		Positive schema Negative schema		
	Not-feminine (tomboy)		Not-masculine (effeminate)	
		Not-feminine not-masculine abstract/angelic		

Figure 1.5. *Example of a semiotic square for feminine/masculine*

Finally, as opposition is essentially a relation between signifieds, it is naturally affected by polysemy. The same lexical item can have several antonyms according to its different significations.

1.1.2.7. *Paronymy*

This is a relationship between two or more words that resemble each other phonetically and/or in terms of spelling without necessarily having a semantic relation. There are several cases of this type, including: *affect* (act physically on something) and *effect* (a phenomenon that follows and is caused by some previous phenomenon); *desert* (arid land) and *dessert* (a dish), and *just* and *justice*. Note that some paronyms are common sources of errors in language use, as in the pair: *diffuse* (to spread) and *defuse* (reduce danger or tension).

1.1.2.8. *Troponymy*

Initially proposed by [FEL 90a], this relation pertains to the classification of verbs regarding the genericity of the events they express: *communicate* > *speak* > *whisper* > *yell*. These relations can be expressed as the *manner_of*. *Poisoning* is a manner of *killing* and *running* is a manner of *moving*.

1.1.3. *Theories of lexical meaning*

1.1.3.1. *The Aristotelian approach*

Categorization is a fundamental cognitive process. It is a means of grouping different entities under the same label or category. Studying this process necessarily involves a good understanding of the nature of categories that humans use. Are they objective categories that depend on the nature of the objects that they represent? Or, on the contrary, are they subjective categories whose existence depends on a community of agents[4]? Two approaches attempt to shed light on these categories: the Aristotelian approach and the prototype approach.

The Aristotelian approach, sometimes called the classical approach or the necessary and sufficient conditions approach, stipulates that the properties shared by the entities in question are the basis of the grouping. So, to determine whether a given entity belongs to a given category, it must possess the properties of this category (necessary conditions) and it must possess enough of them to belong to it (sufficient conditions). At the end of this process, a binary decision is made: the element either belongs to the category in question or not. For example, in order for an entity *X* to be classified in the category *man*, the following conditions must be met:

– X is a human

– X is a male (sex/gender)

– X is an adult

If at least one of these conditions is not met, then *X* is not a man because the conditions are necessary individually. Inversely, we can also logically infer all of the conditions starting from the category: from the sentence *John is a man*, we can infer that *John is a human, John is a male* and *John is an adult*. If all of these conditions are satisfied, then the entity *X* is categorized as a man because all of these conditions are sufficient for the categorization. In other words, from these criteria, we can infer that he is a man and nothing else.

4 For a discussion of the differences between these two points of view, see [PAC 91].

The Aristotelian approach, where the borders between categories are rigid, has been challenged by several works in philosophy and psychology. In his philosophical investigations, the Austrian philosopher Ludwig Wittgenstein from the analytical school showed that it is impossible to define a concept as banal as a *game* in terms of necessary and sufficient conditions. There is at least one case where the following intuitive conditions are not valid:

– Involves physical activity: chess and video games are well-known examples of non-physical games.

– There is always a winner and a loser: many video games are designed in terms of steps and/or points. The notions of victory and loss are not relevant in these cases.

– For leisure: there are professional sports players.

Wittgenstein concluded that categories ought to be described in terms of similarities between families. In these cases, the members of the family are all similar without necessarily sharing common features. In other words, the connections between the members of a given category resemble a chain where the shared features are observed at a local level. Another philosopher from the analytic tradition, the American Hilary Putnam, proposed a similar model known as the semantics of the stereotype.

1.1.3.2. *Semic or componential approach*

To represent the lexical meaning, several linguists, starting with the Dane Louis Hjelmslev [HJE 43], have adopted in various forms a componential analysis of the meaning of words using features similar to those used in phonology. As emphasized in [KLE 90], the features have a role similar to that of the necessary and sufficient conditions of the Aristotelian approach.

Bernard Pottier, main defender of componential analysis in France, gave an example of this kind of analysis, which is presented in Table 1.6 [POT 64].

Semes Words	For sitting	Rigid material	For one person	Has feet	With backrest	With arms
Seat	+	-	-	-	-	-
Chair	+	+	+	+	+	-
Armchair	+	+	+	+	+	+
Stool	+	+	+	+	-	-
Sofa	+	+	-	+	+	-
Pouffe	+	-	+	-	-	-

Table 1.6. *A semic analysis of the field of the chair according to Pottier*

In the example given in Table 1.6, each line represents a sememe. This consists of the set of semes in a word. The seme of the first column, *for sitting*, is shared by all of the words in the table. Pottier proposed calling *classemes* the group of semes that, as the seme *for sitting*, are used to characterize the class. The sememe of the word *seat* is the least restrictive: only the classeme is required because the word is the hyperonym of all other words in the table.

Situated at a more general level, the representation of lexical information quickly becomes more complex. The class of seat itself belongs to the more general class of furniture. In turn, furniture belongs to higher classes such as manufactured objects, and objects in general. Similarly, the class of armchairs includes several subclasses such as the wing chair, Voltaire chair and club chair that each has a set of semes that distinguish them from the set of neighboring or encompassing classes (hyperonyms). This means that a large number of new semes must be added to the semes identified by Pottier himself in order to account for these relations. The word seat itself can be employed figuratively in ways that are different from the ordinary usage, such as *the seat of UNESCO is in Paris*. To account for these uses, Pottier admitted the existence of particular semes, called virtuemes, that are activated in particular cases.

As highlighted in [CRU 00], the principle of compositionality is far from universal. There are phenomena that are an exception to this principle. This includes fixed expressions like *kicked the bucket*, *a piece of cake* and *porte-manteau* as well as the metaphors *the ball is in John's court, to weave a tangled web* and *to perform without a safety net*. There are no objective rules that make it possible to decide which features should be included in a linguistic description. The amount of detail in the descriptions often depends on the specific objectives of each project. This considerably limits the reusability of these works. This point is all the more problematic because the practical implementation of them requires a considerable amount of work.

1.1.3.3. *Prototype semantics*

The ideas of Wittgenstein and Putnam were taken up and developed by the American psychologist Eleanor Rosch and her collaborators who proposed the prototype-based approach commonly called prototype semantics [ROS 73, ROS 75, ROS 78, KLE 90, DUB 91]. According to Rosch, categorization is a structured process that is based on two principles: the principle of cognitive economy and the principle of the structure of the perceived world.

According to the principle of cognitive economy, humans attempt to gain the maximum possible information about their environment while keeping their cognitive efforts and resources to a minimum. Categories serve to group different entities or stimuli under a single label contributing to the economy of the cognitive representation.

The principle of the structure of the perceived world stipulates that the world has correlational structures. For example, *carnivore* is more often associated with teeth than with the possibility of living in a particular zone like the tropics or the North Pole. Structures of this type are used to form and organize categories.

These two principles are the basis for a cognitive system of categorization that has a double dimension: a vertical dimension and a horizontal dimension.

The vertical dimension emerged from Rosch's work concerning the level of inclusion of objects in a hierarchy of categories connected by a relation of inclusion [ROS 76]. For example, the category *mammal* is more inclusive than the category *cat* because it includes, among others, entities such as *dog*, *whale* and *monkey*. Similarly, the category *cat*, which includes several breeds is more inclusive than *Chartreux* or *Angora*.

According to Rosch, the levels of inclusion or abstraction are not cognitively equivalent. The experiments that she conducted with her collaborators showed that there is a level of inclusion that best satisfies the principle of cognitive economy. This level of inclusion is called the base level. It is located at a middle level of details between, on the one hand, a higher level like *mammal* and *vehicle* and, on the other hand, a subordinate level like *Chartreux* and *sedan*. Her work also showed that this base level has several properties that make it cognitively prominent. Among others, they showed that the base level is one where the subjects are most at ease providing attributes. They also showed that the words corresponding to the base level are the first to emerge in the vocabulary, thus proving their primacy in the process of acquisition. Rosch and her collaborators considered that the primacy of the base level affected the very structure of language because we can observe that the words corresponding to the base level are generally simple and monolexical like *chair*, *dog* and *car* contrary to words on subordinate levels that tend to be compound or polylexical words like *key chain, Swiss Army Knife* and *lounge chair*. Finally, they showed that the words in the base level are more commonly used than those in the superordered and subordinate levels. Rosch went so far as to suggest that in the course of the process of a language's evolution, the base-level words emerged before the words in the other two levels [ROS 78].

The horizontal dimension, in turn, is fundamentally linked to the principle of the structure of the perceived world It notably concerns the way in which categories reflect the structure of the world. This correlation is maximized when it pertains to a prototype that is the best example of a category. The prototype serves as the fulcrum of the category. Whether other entities belong to the category in question is determined in terms of similarity to the prototype. In other words, the entities in a given category can be central or peripheral depending on their degree of resemblance to the prototype: there are no features that are necessarily shared by all members of a category. This is the effect of typicality. For instance, this leads us to consider that *apple* is the best example of the category of fruit and to consider *olive* as a peripheral case of the same category. Similarly, *sparrow* is the best example of the category of bird, while *ostrich* and *kiwi* are peripheral examples (see Table 1.7 for a comparison of the properties of these two birds).

In Table 1.7, the ostrich differs from the prototype (sparrow) in six points, whereas the kiwi differs in eight points.

It should also be noted that there are individual differences of classification of entities within a set of linguistically and culturally homogenous subjects. In other words, there is no universally homogenous classification.

Attribute	Sparrow	Ostrich[5]	Kiwi
Lays eggs	Yes	Yes	Yes
Has a beak	Yes	Yes	Yes
Covered in feathers	Yes	Yes	Yes
Has short legs	Yes	No	Yes
Has a tail	Yes	Yes	Almost non-existent tail
Small size	Yes	No	Medium
Has wings	Yes	Atrophied wings	Atrophied wings
Can fly	Yes	No	No
Nostrils located at base of beak	Yes	Yes	No (nostrils are at the end)
Seeks food with its eyes	Yes	Yes	No
Diurnal	Yes	Yes	No
Moves on the ground by hopping	Yes	No	No
Chirps/sings	Yes	No	Yes

Table 1.7. *Comparison of the attributes of sparrows, ostriches and kiwis*

Several critiques have been made about prototype semantics (see [LAU 99] for an overview). One of these critiques is the lack of prototype in certain cases where it is not possible to describe a prototype. For example, *the president of Spain* is a category that does not exist, and is therefore impossible to describe using a prototype, even though it has meaning. Another problem is ignorance or error because it is not able to explain how to address a concept while having an erroneous

5 The features considered for the kiwi and the ostrich were taken from articles corresponding to these two birds in the Encyclopedia Encarta DVD [ENC 09].

understanding of some of its properties. To illustrate this idea, [LAU 99] gives the example of the prototypical grandmother, often described as an old woman with gray hair who wears glasses. This prototype can produce an error by leading us to interpret all women with these features as grandmothers. Inversely, the prototype can lead us to incorrectly exclude cases. For example, a cat remains a cat even without some of its prototypical features (such as the tail, whiskers or ears).

The applications to lexical semantics remain the most important where they pass from the best example of a category to the best use of a lexical unit (for example, see [KLE 90, FUC 91, MAR 91, RAS 91a]). Applied to syntax, the prototype theory makes it possible to distinguish between prototypical uses of syntactically ambiguous words that correspond to several grammatical categories [CRO 93]. As emphasized in [KLE 90], the concept of the prototype also has interesting applications in phonology, morphology and textual linguistics.

1.1.3.4. *The generative lexicon theory*

The theory of a generative lexicon is a theory that highlights the distributed nature of compositionality in natural language. It is mainly based on the work of James Pustejovsky [PUS 91, PUS 95], but it has been developed by other linguists such as [BOU 97, BUS 99]. This theory also gave rise to some computer implementations such as the one in [COP 92]. Two main questions are the basis of this theory. The first concerns the unlimited number of contexts in which a word can be used. The second pertains to the independent nature of lexical information concerning common sense knowledge. In this context, the lexical resources are organized into five different levels: the lexical typing structure, the argument structure, the event structure, the qualia structure and the inheritance structure.

The lexical typing structure gives the type of a word located in the context of a language-type system. Similarly, the argument structure describes the lexical predicate in terms of arity, or number of arguments, and types. This structure can be seen as a minimum specification of its semantic relations. The event structure defines the type of events in an expression. Three classes of events are considered: the states e^{et}, the processes e^{p} and the transitions e^{t}. An event e^{T} can be analyzed in two structured sub-events (e^{p}, e^{et}). The qualia structure describes the semantic properties through four roles:

– The formal role concerns the base category that distinguishes the meaning of a word in the context of a larger domain.

– The constitutive role pertains to the relation between the object and its components.

– The telic role concerns the identification of the function of a word.

– The agentive role pertains to the factors involved in the origin.

– Thus, a word such as *car* can receive the following structure in Figure 1.6.

```
Car
argstr [arg1 = x:artifact]

            Const= {engine, wheels, seats, etc.}
            Formal= x
Qualia      Telic= transport (e1, x,y)
            Agentive= make (e1, x, y)
```

Figure 1.6. *Lexical structure of the lexical entry car*

In the example in Figure 1.6, the word *car* is considered to be an artifact. It has a telic role and its primary function is to transport people and/or merchandise. It is important to distinguish between two types: basic types that are defined in argument structure and higher level types (like events). The latter are accessible through generative operations like government and binding. Government is a coercion operation that converts an argument into the type expected by the predicate to avoid an error. Consider *I finished the book*. In this case, the verb *to finish* involves the government of the noun type *book* in an action related to this noun (reading). Binding makes it possible to modify the telic role of a word without changing its denotation. To illustrate this operation, consider these three sentences with different interpretations of the word *fast* [PUS 91][6]:

– A fast highway → where we can drive fast.

$Q_T(\text{highway}) = \lambda x \lambda e^P[\text{travel(cars)} \wedge (e^P)\ \text{in}(x)(\text{cars})(e^P) \wedge \text{fast}\ (e^P)]]$.

– A fast typist → who types fast.

$Q_T(\text{typist}) = \lambda x \lambda e^P[\text{type}(x)(e^P) \wedge \text{fast}\ (e^P)]$.

– A fast car → that goes fast.

$Q_T(\text{car}) = \lambda x \lambda y \lambda e^P[\text{goes}(x)(y)(e^P) \wedge \text{fast}(e^P)]$.

6 See section 2.2.3 for a presentation of the syntax of lambda expressions.

These interpretations are all derived from the meaning of the word *fast* whose semantics modify the telic role of the noun. For example, in the case of the *fast highway*, it gives the following result:

$$\lambda x \text{ [highway } (x) \ldots \text{ [Telic}(x) =$$
$$\lambda e^P \text{ [travel (cars) } (e^P) \wedge \text{in}(x)(\text{cars})(e^P) \wedge \text{fast}(e^P) \text{]]]}$$

Finally, the inheritance structure indicates how a word is related to other concepts in the context of the lexicon. Pustejovsky distinguishes two types of heritage: fixed and projective. Fixed inheritance includes inheritance methods similar to those used in artificial intelligence (for example, see [BOB 77]). To discover the relationships between concepts like hyponymy and hyperonymy, a fixed diagram must be used. The projective inheritance proposed by Pustejovsky operates in a generative way starting from the qualia structures which are intimately related to the idea of the prototype. To illustrate the difference between these two types, Pustejovsky proposes these two examples [1.4]:

The prisoner escaped last night.
The prisoner ate supper last night. [1.4]

In examples [1.4], the relation between *prisoner* and the action of escaping is more direct than the relation with verbs expressing ordinary actions like eating or sleeping.

1.2. Lexical databases

The first known bilingual dictionary was created in the kingdom of Ebla in what is now north-west Syria, close to the city of Aleppo, in the year 2300 B.C.E. It was a Sumerian-Akkadian dictionary carved onto clay tablets. Other archeological discoveries have brought to light other dictionaries in Babylon (around 2000 B.C.E.) and later in China (in the second century B.C.E.) (see the entry *Dictionary*[7] on Wikipedia for more details). This indicates an interest in creating dictionaries since the dawn of human civilization. With important developments in the means for humans to communicate, the interest in such dictionaries was even more accentuated.

Several automatic natural language processing applications use structured lexical resources in the treatment process. Often created in the form of some kind of database (relational or not, structured or semi-structured), these resources are intended to provide easy access to information related to words, especially their morphology and semantics. The entries in a lexical database can contain other

7 http://en.wikipedia.org/wiki/Dictionary

information depending on the linguistic theory adopted. The quality of a lexical database is determined based on criteria like the following:

– Description of words: the linguistic description of words must be as complete as possible. Thus, all relevant linguistic features must be included. The problem is that linguistic applications vary in terms of requirements. For example, databases destined for a superficial analysis or information search applications require less information about the morphology of a word than a spelling and grammar corrector.

– Dataset coverage: it is generally accepted that it is not possible to include all of the words in a given language in a database, regardless of its size. However, depending on their needs, databases differ in terms of coverage. Some have fairly modest objectives such as the coverage of a specific task, like in task-oriented human–machine dialogue systems. Others, such as the ones used by generic automatic translation systems, tend to be as large as possible.

– Flexibility: it should not be complicated or costly to modify the structure or the content. In particular, it must be easy to add new entries to the base to adapt to the constant evolution of the vocabulary of a language.

– Portability: the database must be compatible with the maximum number of platforms and programming languages to maximize its use by the community.

– Ease of access to information: the database must be easily readable by both humans and machines. Humans need to access the database to write and test grammar, maintain the database, etc. Access to the database through a computer program must also be facilitated in order to reduce the maximum amount of research time for a word and guarantee the quality of the results.

When talking about electronic or paper dictionaries, two concepts should be addressed: the macrostructure and the microstructure. The macrostructure concerns the number of lexical entries covered by the dictionary. Generally, 40,000 entries is considered an acceptable number. The macrostructure also concerns the angle through which the entries are presented: semasiology or onomasiology. A semasiologic approach starts from the word to find the meaning. This approach is used by dictionaries like the Petit Robert [ROB 67]. The onomasiologic approach is related to the semantic content and it is used by dictionaries like the Petit Larousse [AUG 22].

The microstructure concerns the structure and content of the entry. Lexical entries from one dictionary to another are distinguished by very varied information. This includes information such as the social connotation of a word such as formal or informal, morphological information such as the plural or feminine form of a word, etymological information about the origin of a word and the pronunciation in the form of a transcription in the International Phonetic Alphabet or a sound file in the case of electronic dictionaries.

Because databases are, in the end, only a set of electronic documents with particular relationships between them, it is useful to discuss electronic document standards before addressing lexical databases properly speaking.

1.2.1. *Standards for encoding and exchanging data*

Because a lexical database or an electronic dictionary is a collection of electronic documents, it is important to understand the main standards currently available to encode these documents and the standard formats to exchange them. As we will see in sections 1.2.3 and 1.2.4, the content standards as well as the writing systems of dictionaries are closely connected with the standards for encoding and exchanging data.

1.2.2. *Standard character encoding*

To encode information in a database in American English, the *American Standard Code for Information Interchange* (ASCII)[8] was proposed in 1968. It associates digital codes from 0 to 127 with 8-bit characters. For example, the lowercase letter *a* is associated with the code 97 and the character } is associated with the code 125. With the popularization of computers beyond the United States during the 1980s, the need for a multilingual standard began to make itself felt. This led to the creation of the *Unicode Transformation Format* (UTF). At the start, the size of the characters was 16 bits for this standard, but, to include new languages, it was enlarged to 31 bits, thus allowing for more than two billion characters. To reduce the disadvantages related to its large size, a compressed version of this format was proposed. This is the UTF-8 format whose main properties include:

– All code points of the Unicode can be represented.

– A sequence of ASCII characters is also a valid UTF-8 sequence.

– It makes it possible to use languages like Arabic, Korean and Chinese.

1.2.2.1. *SGML*

Standard Generalized Markup Language or SGML is a markup language that became an international standard to define the structure of electronic documents in

8 http://www.asciitable.com/

1986. It is commonly used by publishing houses, which explains its adoption by several dictionaries.

SGML is a metalanguage. This means that it is designed to specify languages. Consider the SGML document shown in Figure 1.7 as an example.

```
<week>
    <day num=1>Monday
    .....
    <weekend>
            <day num=6>Saturday
            <day num=7> Sunday
</week>
```

Figure 1.7. *Extract of an SGML document that represents the days of the week*

In Figure 1.7, the document is delineated by the two *week* tags. The first one is called the start tag and the second one is called the end tag. The end of an element is not systematically marked by an end tag. Indeed, the simple addition of a start tag of the same type as above is considered enough to mark the end of the previous element. For example, adding the start tag <day num=7> also marks the end of the element <day num=6>.

To understand the role of SGML as a metalanguage, consider Figure 1.8. The logic makes it possible to define generic types of documents, such as a monolingual or bilingual dictionary, which in turn serves as a model to construct real documents.

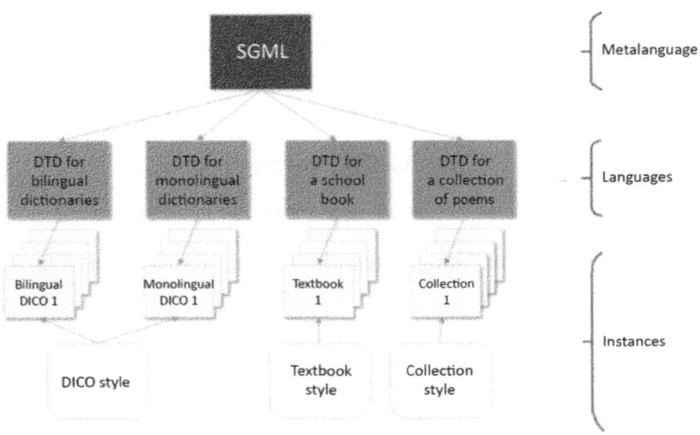

Figure 1.8. *Diagram of a possible use of SGML in a real context*

According to the standard ISO 8879 implemented in 1986, two main levels are distinguished within SGML:

– The logic that is declared in the *Document Type Definition* (DTD). The DTD plays the role of grammar for a type of document whose structure it describes. Thus, its role is to detect anomalies in a document or to help to determine compatibility with the logic. To do this, a DTD must indicate the names of the elements, the nature of the content of each of the elements, the order of appearance of the elements and the authorized frequency of each element (e.g. a book only has one title, but it has several chapters), the possible attributes, and the default values. To understand what a DTD is, consider the structure of a lexical entry in a monolingual dictionary given in Figure 1.9. In this example, a lexical entry is composed of an entry, a gender, a plural form, a phonetic transcription, an etymology and one or more meanings. Each meaning is related to an explanation as well as an example. All of the data retained in this dictionary are textual and can contain SGML tags that are recognized as such (PCDATA)[9].

– Instances are documents realized according to the restrictions expressed in the DTDs. The form of display will be rendered at the end by stylesheets. Stylesheets can be associated with one or more documents at a time.

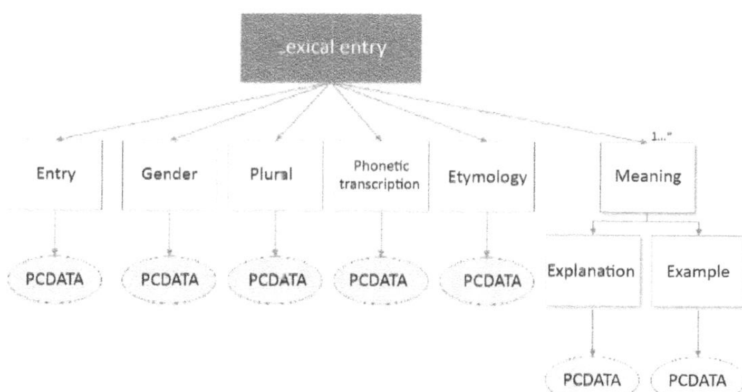

Figure 1.9. *Structure of a lexical entry*

To transform the diagram given in Figure 1.9, the DTD language that has an expressive power similar to regular expressions is used. A complete tutorial about DTDs is beyond the objectives of the present section, but any reader who wishes to learn more can consult Victor Sandoval's book [SAN 94].

9 Unlike CDATA data, which are textual fields in which SGML tags simply have their character sequence values.

The structure of an SGML document is composed of three parts:

– The SGML declaration that defines the adopted characters coding scheme.

– The prologue that contains the DOCTYPE declaration and the reference DTD for the document.

– The instance of the document that contains at least one root element and its content.

To make the presentation more concrete, consider the document shown in Figure 1.10. This document includes the definition of a word in SGML format, while respecting the DTD corresponding to the structure given in Figure 1.10. In addition to the declaration and the prologue, this document includes the following elements: entry (faïence), gender (feminine), plural (faïences), phonetic transcription, etymology (from Faenza, a city in Italy) and a list of two meanings. Each of these meanings has an explanation and an example.

```
<!SGML "ISO 8879:1986"
     -- Basic SGML declaration --
  CHARSET  BASESET   "ISO 646:1983//CHARSET International
                      Reference Version (IRV)//ESC 2/5 4/0"
 >
<!DOCTYPE bdlex -- prologue --SYSTEM "bdlex.dtd">
<lexical_entry meaning="2" >
<!-- definition of the word faïence -->
<entry>
        <entry>faïence</entry>
        <gender>Gender: feminine</gender>
        <plural>Plural: faïences</plural>
        <tr-phon>Phonetic transcription: fajɑ̃s</tr-phon>
        <etymology> Etymology: from Faenza, city in Italy</etymology>
        <meaning-list>
          <meaning>
                <explanation>glazed pottery object</explanation>
                <example>The archeologist found ancient faïences in old
                Lyon</example>
          </meaning>
          <meaning>
                <explanation>modeling method for clay </explanation>
                <example>a service of faïence</example>
          </meaning>
        <meaning-list>
</entry>
```

Figure 1.10. *Example of the definition of a lexical entry in the form of an SGML document*

An SGML document like the one in Figure 1.10 is difficult for humans to read. As such, it is necessary to convert it into a format that is accessible for humans. To do this, stylesheet types such as CSS (Cascaded Stylesheet) are often used. After its transformation by a CSS, the document in Figure 1.10 can be displayed in the way presented in Figure 1.11. Naturally, the same SGML document can be viewed differently when it is associated with different stylesheets.

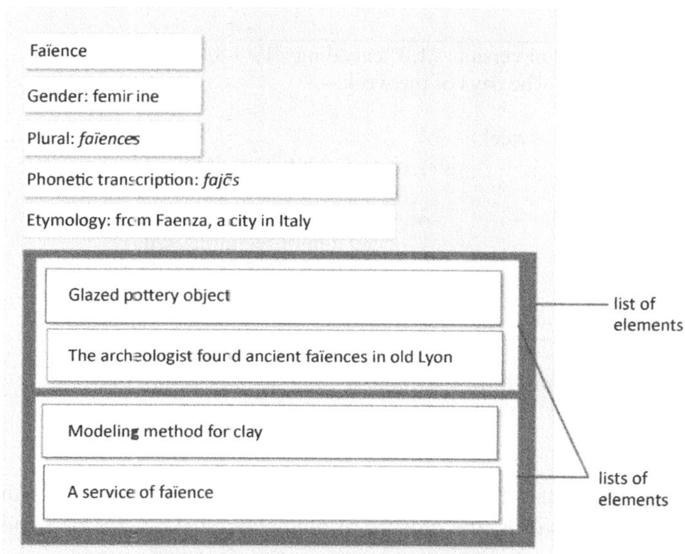

Figure 1.11. *Possible display for the SGML document*

Despite its interest, SGML is not adapted for all uses. In particular, it is not adapted for hyperdocument management, because it was initially only designed to represent the structure of technical documents at a time when the notion of the hyperlink was still in the exploratory stage. This limits the possibilities for Web applications, as opposed to the HTML language, which was specially designed for connecting and reusing documents on the Web.

1.2.2.2. *XML*

Proposed in 1997 as a simplified form of SGML, the e**X**tensible **M**arkup **L**anguage (XML) made it possible to resolve many of the issues related to the unwieldiness of the processing algorithms of SGML documents (for an introduction to XML, see [MIC 01, RAY 03, BRI 10]). Contrary to SGML documents, XML documents have an arborescent structure with only one root element. Moreover, compared with SGML which only defines the concept of the validity of a document (in relation to a DTD), XML also introduces the notion of a "well-formed" document. This new concept allows users to exchange parts of documents and verify that the markup is syntactically correct without needing to know the DTD. Consider the example provided in Figure 1.12.

```
<?xml version="1.0" encoding="ISO-8859-1"?>
<!-- The days of the week -->

    <week>
            <day num=1>Monday</day>
             ...
             <weekend>
                  <day num=6>Saturday</day>
                  <day num=7> Sunday</day>
             </weekend>
    </week>
```

Figure 1.12. *Example of an XML document that represents the days of the week*

All of the elements must start and end with a start tag and an end tag, respectively. In other words, the closure of elements must always be explicit, unlike in SML syntax.

Several satellite languages are closely linked to XML, including:

– DTD: to automatically verify if an XML document conforms to the previously designed format, the diagram DTD (Document Type Definition) is necessary. Alternatives to DTD also exist, such as W3C and Relax NG[10] diagrams.

– The namespaces: these make it possible to include elements and attributes taken from other vocabularies without collision in the same document.

– XML base: this defines the attribute xml:base that resolves relative URI (Uniform Resource Identifier) references in the framework of a document.

– XPath: XPath expressions make it possible to trace the components of an XML document (elements, attributes, etc.).

10 http://www.relaxng.org/

– XSLT: is a language intended to transform XML documents into other formats like XML, HTML and RTF (Rich Text Format). This language is closely linked to XPath that it uses to find the components of the XML document to be transformed.

– XQuery: strongly connected to XPath, XQuery is a query language in XML databases that allows access to and manipulation of data stored in XML documents.

– XSL-FO: this language is mainly used to generate PDF documents from XML documents.

An example of the use of the XML language is the exchange and viewing of data. A possible scenario is presented in Figure 1.13.

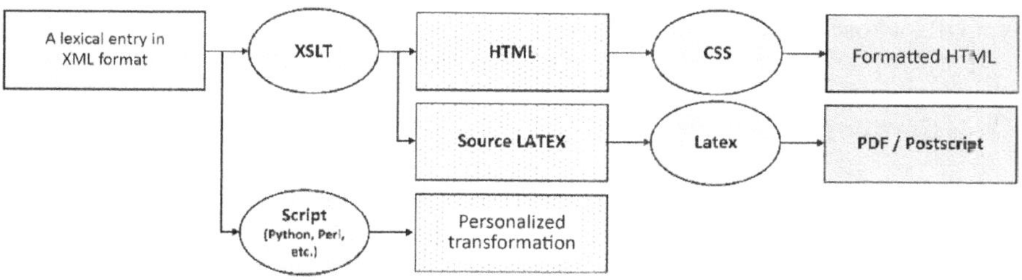

Figure 1.13. *Use of XML to format lexical entries*

Figure 1.13 shows how to transform the XML dictionary entry into various formats using XSLT language or any scripting language like Python or Perl. Beyond the simple viewing of data by human users, this process is useful for exchanging data between several computer programs.

1.2.2.3. *RDF*

Intended for metadata sharing within a community, RDF (Resource Description Framework) provides both a model and a syntax. For a detailed introduction to this language, see [POW 03].

In the RDF model, the concept of the node or data plays an important role. Nodes can be any web resource that has a Uniform Resource Identifier (URI), like a web page or a server. Nodes possess attributes that can have atomic values such as numbers or character sequences, resources or metadata instances. As an example, RDF was adopted to code lexical data extracted automatically from the multilingual dictionary *Wiktionnaire* and make them accessible to the community while guaranteeing their interoperability [SÉR 13].

To illustrate the principles of RDF metadata, see the example given in Figure 1.14.

```
<RDF:RDF>
        <RDF:Description RDF:HREF = "http://URI-of-Document">
        <DC:Creator>John Martin</DC:Creator>
        </RDF:Description>
</RDF:RDF>
```

Figure 1.14. *An example of RDF metadata*

Figure 1.14 shows how the attribute *creator* is attached to a resource identified by a URI whose value is *John Martin*.

1.2.3. *Content standards*

It is generally accepted that a dictionary is an information-rich document. This information can be of various natures and types: morphology, etymology, phonetic transcription, etc. The question that is raised now is how to include information as rich and varied as what is contained in a dictionary, while guaranteeing the interoperability of the dictionary created. To answer this question, we will review three standards that have been determined to be representative to present dictionary content: TEI, SALT and LMF.

1.2.3.1. *The TEI[11] standard*

The *Text Encoding Initiative* (TEI) is an international standard for publishers, museums and academic institutions. This standard intended to develop directives to prepare and exchange electronic material. It was developed between 1994 and 2000 by several groups of researchers who received funding from several institutions such as the European Union, the National Endowment for the Humanities (United States) and the Canadian National Research Council [IDE 95, JOH 95]. Several DTDs were proposed following this project for several types of texts: prose, poetry, dialogues and different types of dictionaries.

Two parts can be distinguished within the TEI standard: a discursive description of texts and a set of tag definitions. These definitions serve to automatically generate frames in several electronic formats such as DTD and RELAX NG.

---

11 http://www.tei-c.org/

In a TEI dictionary, the information is organized like this:

– the form of the word: spelling, pronunciation, accentuation, etc.;

– the grammatical form: categories, sub-categories, etc.;

– the definition of the word or its translation;

– the etymology of the word;

– links;

– similar entries;

– information about its use.

Consider the example of the entry *dresser* encoded following the TEI format in Figure 1.15. This entry contains several types of information including morphological information, semantic information (domain, synonym), translations, examples, etc.

```
<entry n="1">
<form>
		<orth>dresser</orth>
</form>
<sense n="a">
<sense>
<usg type="dom">Theat</usg>
<cit type="translation" xml:lang="fr">
		<quote>habilleur</quote>
		<gramGrp>
		<gen>m</gen>
		</gramGrp>
</cit>
<cit type="translation" xml:lang="fr">
		<quote>-euse</quote>
		<gramGrp>
		<gen>f</gen>
		</gramGrp>
</cit>
</sense>
<sense>
		<usg type="dom">Comm</usg>
		<form type="compound">
		<orth>window <oRef/>
		</orth>
		</form>
		<cit type="translation" xml:lang="fr">
		<quote>étalagiste</quote>
		<gramGrp>
```

```
            <gen>mf</gen>
            </gramGrp>
            </cit>
</sense>
<cit type="example">
            <quote>she's a stylish <oRef/>
            </quote>
<cit type="translation" xml:lang="fr">
            <quote>elle s'habille avec chic</quote>
</cit>
</cit>
<xr type="see">V. <ref target="#hair">hair</ref>
</xr>
</sense>
<sense n="b">
            <usg type="category">tool</usg>
<sense>
<usg type="hint">for wood</usg>
<cit type="translation" xml:lang="fr">
            <quote>raboteuse</quote>
            <gramGrp>
            <gen>f</gen>
            </gramGrp>
</cit>
            </sense>
            <sense>
            <usg type="hint">for stone</usg>
            <cit type="translation" xml:lang="fr">
            <quote>rabotin</quote>
            <gramGrp>
            <gen>m</gen>
            </gramGrp>
</cit>
</sense>
</sense>
</entry>
<!-- ... -->
<entry xml:id="hair">
<sense> <!-- ... --></sense>
</entry>
```

Figure 1.15. *Example of the entry dresser [BUR 15]*

TEI has not succeeded in specifying a single standard for all types of dictionaries. However, this standard is doubly interesting. On the one hand, it has succeeded in unifying the SGML tags and, on the other hand, it has specified the semantic content of dictionaries by clarifying concepts such as category, etymology and translation.

Several thousands of books, articles and even poems have been encoded with TEI-XML, a large part of which are currently available for free on the web. As the DTD is very large, a more easily accessible version known as TEI lite has also been proposed.

1.2.3.2. *The SALT project*

Jointly funded by the European Union and the National Science Foundation (NSF) between 1999 and 2001, the *Standards-based Access to multilingual Lexicons and Terminologies* (SALT) project intended to integrate resources used in automatic translation (lexical databases) and terminological data employed in the domain of computer-assisted translation (concept-oriented terminological databases) in a unified framework [MEL 99]. This was a free project, in the software sense of the word, which aimed for the creation of free standards. To do this, the project adopted the XML language as a framework and notably XLT (*eXchange format for Lex/Termdata*). This project aimed to accomplish several tasks:

– Test and refine the data exchange format.

– Develop a website to test the XLT format.

– Develop tools to facilitate the realization of applications with data in XLT format.

Two data exchange formats are combined in the context of the SALT project: the OLIF format (Open Lexicon Interchange Format) and the MRTIF language (Machine-Readable Terminology Interchange Format). The OLIF format concerns the exchange of data between the lexical resources of several automatic translation systems while the MARTIF language is designed to facilitate the exchange of terminological resources intended for human use (see the example of a document in MARTIF format in Figure 1.16[12]).

The document is divided into two main parts: a header and the body. The header describes the source and the encoding of the document. The body of the document includes the term's ID (ID67), the term's domain (manufacturing/industry) and the definition of the term in English and Hungarian.

There are many advantages to a standard like SALT, including the rapid insertion of new terms into a database. This is done using an import/export function of XLT sheets to guarantee coherence in documents that are translated or written by several authors. SALT also allows for the synchronization of translations done by machine or manually. It is more and more common, especially in large institutions, to have hybrid translations: manual translations potentially assisted by computer with

12 http://www.ttt.org/oscar/xlt/webtutorial/termdata.htm

automatic translations potentially post-edited manually. This requires the use of unified terminology throughout all tools and reporting possible gaps in the databases (the lack of certain terms in one base or another).

```
<?xml
 version='1.0'?>
<!DOCTYPE martif PUBLIC "ISO 12200:1999A//DTD MARTIF core
(MSCcdV04)//EN">
 <martif type='DXLT' lang='en' >
 <martifHeader>
		<fileDesc><sourceDesc><p>from an Oracle termBase</p></sourceDesc>
		</fileDesc>
		<encodingDesc><p type='DCSName'>MSCdmV04</p></encodingDesc>
</martifHeader>
<text> <body>
		<termEntry id='ID67'>
			<descrip type='subjectField'>manufacturing</descrip>
			<descrip type='definition'>A value between 0 and 1 used in
...</descrip>
			<langSet lang='en'>
			<tig>
				<term>alpha smoothing factor</term>
				<termNote type='termType' >fullForm</termNote>
			</tig>
			</langSet>
			<langSet lang='hu'>
			<tig>
				<term>Alfa simítási tényezõ </term>
			</tig>
			</langSet>
		</termEntry>
 </body> </text>
</martif>
```

Figure 1.16. *Example of a MARTIF format document*

1.2.3.3. *The LMF standard*

LMF or Lexical Markup Framework is the standard ISO 24613 for managing lexical resources. Developed in 2008, it has the following objectives:

– managing lexical resources;

– offering a meta-model for managing lexical information at all levels;

– offering encoding and formatting specifications;

– making it possible to merge several lexical resources;

– covering all natural languages including the ones that have a rich morphology like Arabic or Finnish.

LMF uses Unicode to represent scripts and the spelling of lexical entries. The specification of the LMF standard respects the principles of Unified Modeling Language (UML). Thus, UML diagrams are used to represent structures, while instance diagrams are used to represent the examples. Linguistic categories like Feminine/Masculine and Transitive/Intransitive are specified in the *Data Category Registry*.

As shown in Figure 1.17, the LMF standard includes several components that are grouped into two sets: the node and the extensions [FRA 06]:

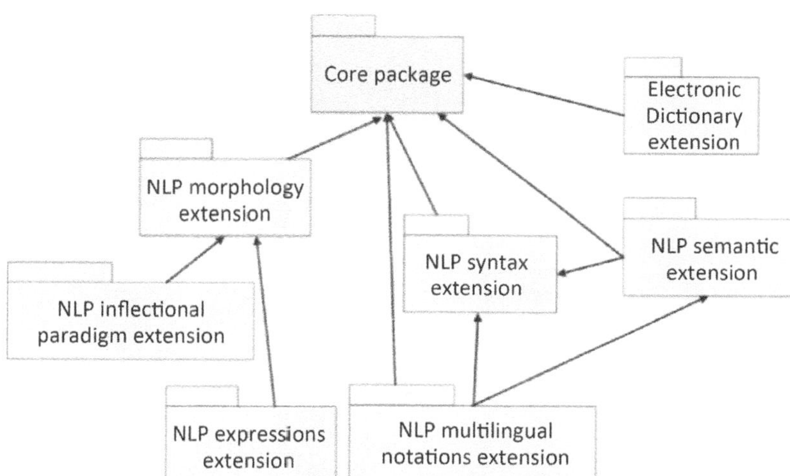

Figure 1.17. *The components of the LMF standard*

The extensions are described in the appendices of the document ISO 24613 as UML packages. They include an electronic dictionary as well as lexicons for NLP. If needed, a subset of these extension packages can be selected, although the node is always required. Note that all of the extensions are compatible with the model described by the node, to the extent that certain classes are enriched by extension packages.

The node whose class diagram is presented in Figure 1.18 describes, among other things, the basic hierarchy of the information in a lexical entry.

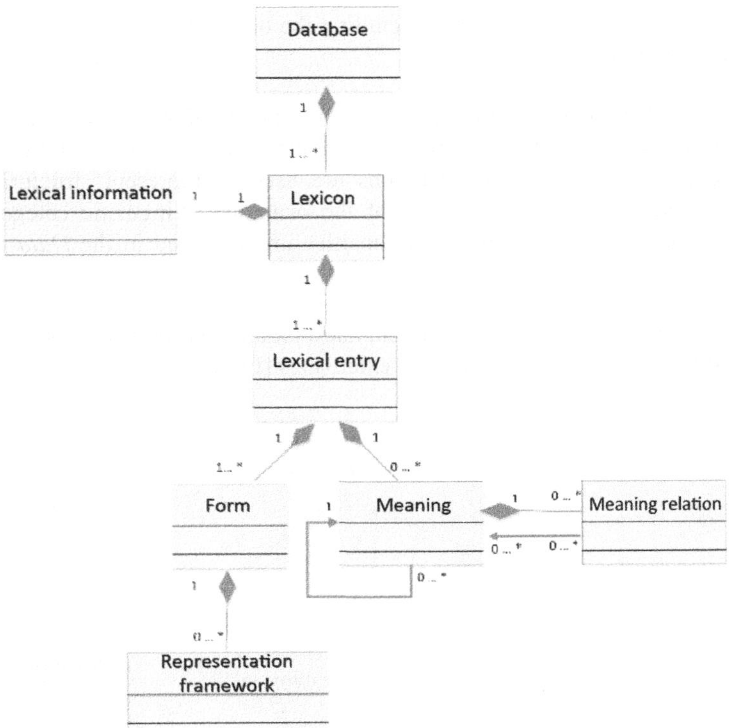

Figure 1.18. *Class diagram of the LMF's core*

As shown in Figure 1.18, the database is composed of an undefined number of lexicons. Composed of an undefined number of lexical entries in turn, each lexicon is associated with some lexical information. Each lexical entry has a relation of composition with one or more meanings as well as one or more forms.

Consider the example in Figure 1.19, which represents two WordNet synsets (see section 1.2.5.1). Each gloss is divided into one instance of *SemanticDefinition* and possibly several *statement* instances. The two *synset* instances are also connected by a *SynsetRelation* instance.

```
<LexicalEntry>
        <DC att="partOfSpeech" val="noun"/>
         <Lemma>
                <DC att="writtenForm" val="oak tree"/>
        </Lemma>
        <Sense id="oak_tree0" synset="12100067"/>
</LexicalEntry>
<LexicalEntry>
        <DC att="partOfSpeech" val="noun"/>
        <Lemma>
        <DC att=writtenForm" val="oak"/>
        </Lemma>
        <Sense id="oak0" synset="12100067"/>
        <Sense id="oak2" synset="12100739"/>
</LexicalEntry>
<Synset id="12100067">
        <SemanticDefinition>
        <DC att="text" val="a deciduous tree of the genus Quercus"/>
        <Statement>
                <DC att="text" val="has acorns and lobed leaves"/>
        </Statement>
        <Statement>
                <DC att="text" val="great oaks grow from little acorns"/>
        </Statement>
        </SemanticDefinition>
        <SynsetRelation targets="12100739"
                <DC att="label" val="substanceHolonym"/>
        </SynsetRelation>
</Synset>
        <Synset id="12100739">
        <SemanticDefinition>
        <DC att="text" val="the hard durable wood of any oak"/>
        <Statement>
                <DC att="text" val="used especially for furniture and flooring"/>
        </Statement>
        </SemanticDefinition>
</Synset>
```

Figure 1.19. *Example of a lexicon coded with LMF [FRA 06]*

1.2.4. *Writing systems*

As the process of writing a dictionary or a lexical database is far from being simple, it is increasingly necessary to use advanced tools to accomplish it. Given the considerable developments of new technologies, it is becoming more common to see teams distributed over a large geographical area collaborating on a shared project. This requires tools adapted for this new mode of working to guarantee the integrity and homogeneity of the work.

Several projects were developed to create advanced dictionary writing systems, including: Papillon, DEB, the Longman Dictionary Publishing System DPS, and the TshwanLex. For brevity's sake, this discussion will be limited to the Papillon and DEB projects.

1.2.4.1. *Papillon*

This project intended to create a multilingual database that covers languages as varied as English, French, Japanese, Thai, Chinese and Lao. It consists of an open-source project that is freely accessible for non-commercial uses. Initiated in 2000, it was funded by the French embassy in Japan as well as the National Institute of Informatics (NII) in Japan [SÉR 01, BOI 02, MAN 06]. The initial phase of the project included only three languages (FEJ: French, English and Japanese) and two teams were involved: NII and the GETA team from the CLIPS-IMAG laboratory in Grenoble.

Inspired by the works of Bernard Vauquois on automatic translation, the idea of the macrostructure of a dictionary is based on a central point that connects the monolingual entries to one another. This kind of structure is particularly practical for adding new languages, as there is no need to link all of the entries to their equivalents (see Figure 1.20).

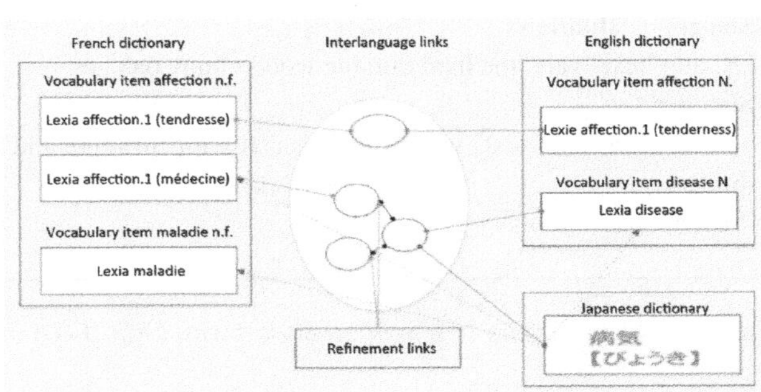

Figure 1.20. *Papillon macrostructure with interlanguage links [MAN 06]*

The macrostructure of the multilingual pivot is based on the PhD thesis work of Gilles Sérasset [SÉR 94]. It consists of a monolingual volume for each language included in the dictionary and one independent pivot volume. The entries in different languages are connected through interlanguage senses. These senses are themselves interconnected by refinement links whose role is to treat the semantic divergences between the languages. Consider the example presented in Figure 1.21.

Axie#456 --lg-->	FR(maladie#1), EN(disease#1)
Axie#457 --lg-->	FR(affection#2), EN(affection#3)
The concatenation of monolingual links gives rise to the following interlanguage links:	
Axie#500 --lg-->	FR(maladie#1, affection#2), EN(disease#1, affection#3)

Figure 1.21. *Examples of interlanguage links [BOI 02]*

In Figure 1.21, the two meanings of the word affection (affection and disease) are both related to a sense in the pivot. In turn, these senses will be connected to other entries in other monolingual dictionaries. As shown in Figure 1.22, the senses are translated into UNL language, which is the pivot representation format [UNL 96].

```
<axi id="a001">
        <lexies>
        <lexy
              lang="fra"
              ressource='papillon-fr.xml'
              idref="meurtre#n.m.@1"/>
   </lexies>
        <external_references>
         <UWs ressource="UNL-fr.unl">
          <uw idref="murder" />
         </UWs>
        </external_references>
</axi>
```

Figure 1.22. *Example of an interlanguage sense in XML [SÉR 01]*

The microstructure of monolingual entries is inspired by the Meaning-Text Theory of Mel'cuk (which will be discussed in section 2.1.5). More specifically, it consists of an adaptation of the lexical database DiCO developed by Alain Polguère at the Université de Montréal [POL 00]. Despite its complexity, this structure was retained because it offers several advantages. On the one hand, it is essentially independent of language. This makes it possible to use the same structure for the different languages included in the project. The very small part of necessarily dependent aspects of the language concerns linguistic properties and register. On the other hand, it was developed for a double usage: use by humans in the context of a classic dictionary and use by machines as a database.

Each lexical unit is made up of a name, its linguistic properties (e.g. part of speech) and a formal semantic definition. In the case of a predicative lexia, the description concerns not only the predicate but also its arguments. A government motif describes the syntactic realization of the arguments and a list of the lexical-semantic functions among the 56 defined by the formalism that are universally applicable to all languages. An example of a lexical entry is given in Figure 1.23.

The lexical entry given in Figure 1.23 shows how the microstructure adopted covers the grammatical properties of the lexical item in question (noun, masculine), the semantic properties (the murder involves an agent and a patient), syntactic dependencies (government relations) that the word involves, lexical functions, an example and idiomatic expressions.

Name of the lexical item: MURDER
Grammatical category: noun.
Semantic formula: action of killing: ~ BY the individual X of the individual Y
Government pattern: X = I = of N, A-Poss Y= II = of N, A-poss
Lexical functions:
{QSYN} assassination, homicide#1, crime /* quasi synonyms */
{Oper1} accomplish, commit, perpetrate [ART ~];
Tremper [in ART ~
{S1} author [of ART ϕ] // murderer-n /* noun for X */
{S2} victim [of ARTϕ] /* noun for Y */
Example: conflict can be a motive for murder.
Idioms:
get away with murder
to scream bloody murder

Figure 1.23. *Microstructure of the lexia murder [MAN 06]*

1.2.4.2. *DEB*

The Dictionary Editor and Browser (DEB) was designed to manage dictionary data, lexical databases, semantic networks and complex ontologies [HOR 07]. It makes it possible to store, index and locate linguistic data. XML is used as the data format and as a means to formalize user interfaces. Note that the structure of the data is flexible, because elements and traits can be added.

The platform is constructed according to a client–server architecture, where clients play a limited role in the graphic or web interface. This allows for some flexibility in the exchange of information, as much for users as for the data interface. Thus, a geographically distributed team can share data easily because the data modifications by one user are seen directly by the other users. To guarantee the integrity of the data, the server is equipped with authentication and authorization tools. In addition, the multiple interfaces offered by the server can be used by different clients at the same time and the programming can be done with any programming language. The DEB adopted the development concepts of the platform Mozilla, whose flagship application is the browser Firefox [FEL 07]. This implies a clear separation between the logic and the definition of the presentation.

Naturally, the use of XML by the server contributes to the interoperability of the data because it is used to develop a large number of various types of dictionaries (monolingual, multilingual, thesauri), semantic networks, ontologies, etc.

The data flow in the DEB system is presented in Figure 1.24.

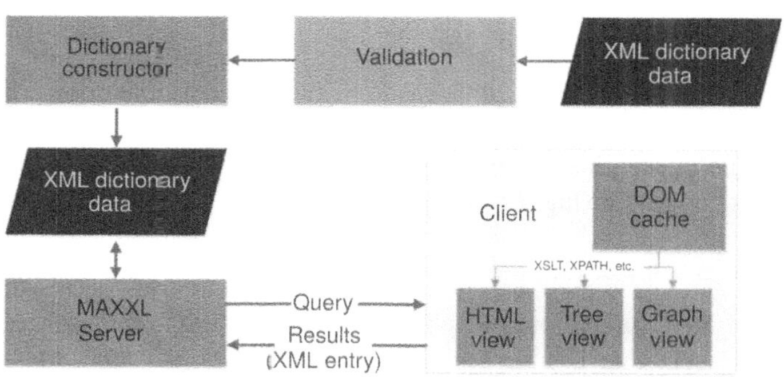

Figure 1.24. *Data flow in the DEB system [SMR 03]*

The server converts XML documents into a binary representation. To concretely explain the function of the server, consider this example of a query. There are two dictionaries: a Czech dictionary and an English dictionary, the Czech WordNet (wn_cz) and an English dictionary defining glosses that will be called gloss_en.

```
<synonym>
        <ili>00004865-n</ili>
        <pos>n</pos>
        <hypernym>00001234-n</hypernym>
        <li sense="1">podvod</li>
        <li sense="1">podraz</li>
        <li sense="1">podfuk</li>
        <li sense="6">bouda</li>
</synonym>
```

Figure 1.25. *Example of a lexical entry in the wn_cz dictionary [SMR 03]*

Figure 1.25 presents a Czech lexical entry that has several meanings. This entry has an identifier (00004865-n) marked by the tag ili that can be linked with an equivalent entry in the English dictionary that shares the same identifier (see Figure 1.26).

```
<en>
        <ili>00004865-n</ili>
        <gloss>an act of deliberate betrayal</gloss>
</en>
```

Figure 1.26. *Example of a lexical entry in the gloss_en dictionary [SMR 03]*

A large number of queries can be made in these two dictionaries. For example, the query wn_cz-* sub "pod" – searches all entries that contain a sub-chain of pod throughout. The query gloss: (wn_cz-li exa "bouda") finds all entries in the wn_cz dictionary that contain the tag *li* with the text *bouda*.

Several projects are associated with the DEB, including:

– DEBDict: this is a dictionary equipped with a multilingual interface, initially in English and Czech, that is able to make queries in several XML databases. The results of these queries can be transformed using the XSLT language. It is also possible to connect with external links such as a morphological analyzer of Czech, sites like *Google* or *answers.com*, or even geographical information systems.

– DEBVisDic: this is a reimplementation of a semantic network editor (VisDic).

– PRALED: this is the preparation for a new exhaustive database for the Czech language (Czech Lexical Database, CLD).

– The visual browser: this is a java application that makes it possible to view coded data according to an RDF diagram. It connects to the DEB server and displays WordNet data.

1.2.5. *A few lexical databases*

There are currently a multitude of lexical resources available for NLP experts or researchers in related disciplines. Intended for a variety of uses in monolingual as well as multilingual contexts (automatic translation, information retrieval, knowledge extraction, etc.), they are distributed in diverse forms like simple lists of words, electronic dictionaries, thesauri, glossaries and databases. We have included a non-exhaustive list of a few resources for different languages. The objective is to give a general idea about these resources and their main advantages. It can also show how the theoretical principles that were discussed in the previous sections are implemented in the form of real databases.

1.2.5.1. *WordNet*

Inspired by work about lexical memory in psycholinguistics, this database was developed at Princeton University in the United States [MIL 90, FEL 05]. A multilingual version of WordNet named EuroWordNet was later developed for European languages [VOS 98]. Other versions for other languages such as French [SAG 08], Arabic [BLA 06], Polish [VET 07] and Romanian [TUF 04] were also created. An extension for a better representation of verbal forms called VerbNet was also proposed by Karin Kipper during her work at the University of Pennsylvania [KIP 05].

WordNet is freely accessible in different forms. For its use as a dictionary intended for human users, WordWeb software[13] developed a simple and practical interface that facilitates navigation and understanding of the structuring of the lexicon that includes around 160,000 roots and 220,000 meanings. In the domain of NLP, several possibilities are offered to programmers. Several Application Programming Interfaces (APIs) are available to integrate WordNet into applications written in Java, C++, C#, etc. WordNet is also integrated within the NLTK tool box in Python, developed at the University of Pennsylvania [BIR 09a]. Widely used, this tool box also offers other tools like morphological analyzer or syntactic parser, chunker, etc.

WordNet was designed to establish connections between four types of parts of speech: nouns, verbs, adjectives and adverbs. The synset is the smallest unit in WordNet. It consists of a structure that represents a particular meaning of a word It

---

13 http://wordweb.info/

includes the word, its explanation, its relations and sometimes one or more use cases. The explanation of the concept of a word is called a *gloss*. For example, the words *night*, *nighttime* and *dark* constitute a single synset in English in this gloss: the time after sunset and before sunrise.

The treatment of polysemy in WordNet occurs by reserving an independent entry for each meaning. The main difference between WordNet and a classic thesaurus is that the unit is not a sequence of characters or a word but rather a meaning. This facilitates the semantic disambiguation of similar words in the network. To clarify this difference, consider the entry *car* presented in Figure 1.27. The result of the search for this word is organized into five synsets, each of which corresponds to one or more synonyms that have the same definition.

S: (n) **car**, A motor vehicle with four wheels; usually propelled by an internal combustion engine *"he needs a car to get to work"*

S: (n) **car**, A wheeled vehicle adapted to the rails of railroad *"three cars had jumped the rails"*

S: (n) **car**, The compartment that is suspended from an airship and that carries personnel and the cargo and the power plant

S: (n) **car**, Where passengers ride up and down *"the car was on the top floor"*

S: (n) **car,** A conveyance for passengers or freight on a cable railway *"they took a cable car to the top of the mountain"*

Figure 1.27. *Result of a search for the word car in WordNet*

The synsets are connected to one another by semantic relations. Among these relations, hyponymy/hyperonymy are the most frequently encoded in the network. It connects generic synsets like *vehicle* to more specific synsets like *car* or *truck*. The root of all the hierarchies is the element *entity*. Among others, WordNet distinguishes between types that concern common nouns and instances that concern specific people, countries or geographic entities. For example, *reptile* is a type of animal while *Mount Rainier* is an instance of a mountain. Always pertaining to concrete entities, the instances systematically correspond to node sheets in the hierarchies.

On a syntactic level, there are four categories in WordNet: nouns, verbs, adjectives and adverbs. Similarly, the content includes the four following units: composite words, idiomatic expressions, collocations and phrasal verbs.

– The nouns, hierarchically organized, are connected according to relations of hyperonymy or hyponymy, coordinating terms that share a common hyperonym like *cat* and *lion*, meronymy as in the relation *part_of* that relates *backrest* and *chair*, or holonymy that expresses the relation *composed_of* that relates *chair* to *backrest*.

– The verbs are organized according to the relations between the activities that they describe. For instance, there are relations like troponomy, which connects verbs where activity *X* described by verb *A* is a sort of activity *Y* described by verb *B*, as with the action *communicate* and the verbs *talk*, *whisper*, etc. [FEL 90b].

– The adjectives are organized in terms of antonymy in the form of pairs of direct antonyms (dry/wet, young/old, hot/cold).

– Since adverbs are directly derived from adjectives in English, this facilitates processing words in this category. WordNet covers relations of synonymy, like oddly/curiously, and antonymy, like between the words quickly/slowly.

1.2.5.2. *The Prolex database of proper nouns*

This project, led by the computing laboratory at the University of Tours, intends to create a platform that includes a multilingual dictionary of proper nouns (Prolexbase), identification systems for proper nouns, local grammars, etc. Two basic concepts are at the foundation of the Prolex model: the pivot and the prolexeme [MAU 08].

Independent of the language, the pivot is constructed on the basis of the notion of quasi-synonyms. These relations have diverse origins:

– Diachronic: historical variations result in changes to names, especially due to changes in political regimes. After the Gulf War and the fall of the Iraqi regime, Saddam City became Sadr City. Similarly, the city of Saint Petersburg was known as Leningrad during the Soviet period.

– Diastratic: well-known people, such as authors, artists, religious figures and political personalities, can sometimes go by different names. So, the following pairs are all quasi-synonyms: (Voltaire, François Marie Arouet), (Apollinaire, Wilhelm Apollinaris de Kostrowitzky) and (John Paul I, Albino Luciani).

– Diaphasic: this process consists of using attractive nicknames to designate some locations like the Golden State for California and the Big Apple for New York City or an indication of the political regime to designate countries in political discourse such as the Kingdom of Belgium and the People's Republic of China.

A prolexeme is the projection of the pivot to a particular language. Three types of dependent variations of the language are the basis for the idea of the prolexeme:

– The name and its written aliases: for example, the United Nations Organization can be designated by a shorter name, the United Nations, or by the acronym UN. Another example consists of using the initials of some political or artistic personalities, such as JFK for John Fitzgerald Kennedy and DSK for Dominique Strauss-Kahn.

– The quasi-synonyms: for example, Caritas USA Organization is a quasi-synonym for the Catholic Relief Service.

– Derivatives of proper names: these are words obtained using a standard morphological process like onus**ian**, onus**ians** and Dickensian. Figure 1.28 presents the pivot 48226 of the prolexeme (UNO, United Nations Organization).

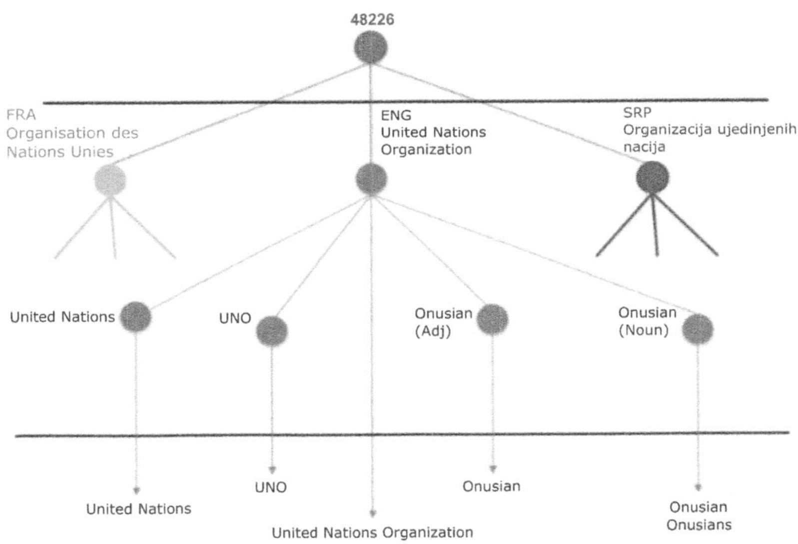

Figure 1.28. *The pivot, prolexeme and instances of the UNO [MAU 08]. For a color version of this figure, see www.iste.co.uk/kurdi/language2.zip*

In Figure 1.28, there are six instances derived from the UNO in English, whose morphology is known to be poor. In French, there are eleven and in Serbian, there are more than fifty.

Finally, it should be noted that such a dictionary is particularly useful for processing journalistic texts, notably the extraction of named entities, where proper names constitute an important part: around 10% of all of the words.

1.2.5.3. *The lexical database Brulex*

Realized between 1988 and 1990 by Alain Content and his collaborators at the Université libre de Bruxelles, Brulex is a lexical database for written and spoken French. The point of departure of this database developed for psycholinguistic research is the Micro-Robert dictionary, which contains 30,000 words.

Brulex provides basic information on each word such as spelling, pronunciation, grammatical class, gender, number and frequency. Information that is useful for selecting experimental material is also provided, notably including the point of uniqueness, counting lexical neighbors, phonological patterns, etc. New specialized resources were added to Brulex, including Lexop [PEE 99] and Manulex [LET 04].

1.2.5.4. *Lexique*

Containing 135,000 French words, this database gradually took the place of Brulex, notably within the psycholinguistic community. It consists of an open database in which the community is encouraged to participate. Regularly updated, this database exists in three versions: Lexique 1, Lexique 2 and Lexique 3 described in [NEW 01, NEW 04, NEW 06], respectively.

Distributed under a license compatible with GNU, Lexique 3 provides a fairly considerable amount of information, the most important parts of which are:

– the gender, number and grammatical category;

– the frequency of words in writing estimated by the Frantext corpus that contains around fifteen million words;

– the syllabic form as well as the number of orthographic neighbors;

– the inflectional family of lemmas and their cumulative frequency;

– the frequency of letters, phonemes, bigrams, trigrams and syllables.

Lexique 3 comes with an online search engine that is called Open Lexique. This tool makes it possible to search seven databases at once (a database of first names, a database of anagrams, a database of orthographic cousins, etc.). Lexique 3 is also equipped with an offline search tool called Undows.

1.3. Knowledge representation and ontologies

1.3.1. *Knowledge representation*

Since the beginning of AI, the use of knowledge necessary for reasoning was most apparent in the context of expert systems. Formalisms to represent knowledge were developed to then create ontologies (for a discussion of these differences, see [GRI 07, SAL 08]).

In the domain of artificial intelligence and automatic natural language processing, knowledge representation is closely linked to the domain of reasoning. Intelligent systems seem to depend significantly on several types of knowledge, including knowledge about the environment. Since this knowledge is typically far from perfect, intelligent systems proceed to operations of deducing new knowledge from present knowledge. Consider the Prolog micro knowledge base in Figure 1.29.

vegetable(bean, green). vegetable(carrot, orange). vegetable(pea, green). vegetable(zucchini, green). vegetable(tomato, red). furniture(chair). furniture (table). furniture (bed). furniture (armoire).	fruit(banana, yellow). fruit(grape, red). fruit(orange, orange). fruit(pear, yellow). fruit(chestnut, brown). edible(X):-vegetable(X, C), write("This vegetable is: "), write(C). edible(X):-fruit(X, C), write("This fruit is: "), write(C).

Figure 1.29. *A program in Prolog with a micro knowledge base*

The names of a few vegetables, fruits and furniture items are stored in the knowledge base in Figure 1.29. The rules of inference make it possible to complete the knowledge by adding facts such as "all of the vegetables are edible", "all of the fruits are edible" and, indirectly, "all of the furniture items are not edible". The concern of these rules for an intelligent or NLP system is that it avoids adding redundant features to the knowledge base, which would weigh it down considerably. This lightness facilitates the modification and maintenance of the base, such as adding a new feature like: "all of the fruits are sweet" or "there are no fruits that are the color black or fuchsia", etc. Naturally, this reasoning only concerns knowledge that is found in the base because it is designed with the hypothesis of a closed world.

1.3.1.1. *Formalisms for knowledge representation*

Generally, all formalisms of knowledge representation and ontologies must allow the expression of the following elements:

– Entities or individuals: these are the basic elements of an ontology. These elements can be concrete like people, vehicles and furniture, or abstract like emotions, numbers and ideas.

– Concepts: this is a means of making collections of objects based on a taxonomy. In other words, concepts make it possible to ground entities or classes of entities. For example, the concept *furniture* includes the class of all furniture items. As emphasized in [BAC 00], it is not possible to identify key concepts or non-logical primitives from which other concepts can be defined, because all concepts are defined in relation to other concepts. Thus, the primitives that are necessary for the formalization and representation of the problem to be solved must be modeled. These primitives must be modeled from a collection of available empirical data: the corpus.

– Attributes: these are elements, often adjectives, that make it possible to describe entities. For example, the entity *Lightning II* (commonly known as F35) has the following attributes: fighter aircraft, single engine, single seat and multimission.

– Relations: these make it possible to model the links between entities or concepts. These relations can play various roles. In some cases, they can play the role of an attribute whose value is another entity, notably in relations of composition (a car is composed of an engine and wheels). They can also express logical, mathematical or chronological relations such as successor(Super Mirage 4000[14], Mirage 2000) and successor(Mirage 2000, Mirage III)). As noted by [BAC 00], a relation is defined in two different ways. On the one hand, it is defined by the concepts that it connects: for example, *to be animated* and *action*. On the other hand, it is defined by the semantic content connecting the two concepts. For example, the action is done by an animated being.

1.3.1.2. *Semantic networks*

Semantic networks consist of representing knowledge in the form of graphs with nodes and arcs. As noted by John Sowa [SOW 92], the oldest known version of a semantic network was proposed by the Neoplatonic philosopher Porphyry of Tyre in his commentaries on the categories of Aristotle. According to Sowa, this version is the ancestor of all modern forms of hierarchies used to define concepts. Near the end of the 19th Century, Charles Peirce proposed the use of so-called existential graphs for knowledge representation. This framework experienced a resurgence in interest toward the end of the 1950s and at the start of the 1960s, especially in the context of automatic translation applications where it was used to represent interlanguage knowledge. The version proposed by Quillian is considered by many researchers as the reference version [QUI 68]. Semantic networks are considered to be a notational alternative to a subset of first-order logic. Contrary to the logic in which notations are sometimes considered rough, semantic networks are distinguished by the ease of displaying knowledge and inferences.

14 Super Mirage is a French jet fighter aircraft, https://en.wikipedia.org/wiki/Dassault_Mirage_4000

From a syntactic perspective, semantic networks are composed of two elements:

– Nodes: represent entities, attributes, events, states, etc. To refer to different individuals of the same type, a different node is used for each individual.

– Arcs: represent the relations between the concepts that they connect. These relations can be linguistic (agent, patient, recipient, etc.), logical, spatial or temporal cases. A label on each arc indicates its type.

An example of a semantic network is represented in Figure 1.30.

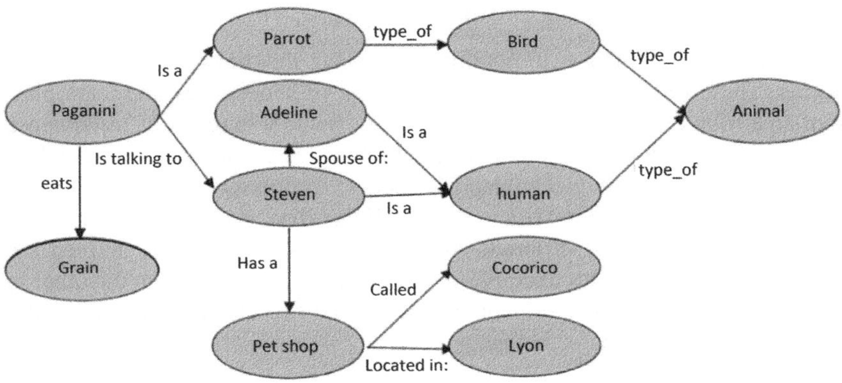

Figure 1.30. *Example of a simple semantic network*

In the semantic network represented in Figure 1.30, taxonomic knowledge (type of) and knowledge related to a particular context (is located at, is called, speaks to, etc.) can be represented in the same network.

Inheritance is one of the key properties of semantic networks. It allows a particular entity to appropriate the features of the class to which it belongs while having specific features. For example, the cats Felix and Fedix inherit all features of the class of cats to which they belong (e.g. the feature *has a tail*) but they also have the specific color features *white* and *black with an orange spot on the nose*, respectively. Exceptional features can also be expressed like the feature *does not have fur*. In the example given in Figure 1.30, birds and humans inherit the features of the category *animal*.

Like some object-oriented programming languages such as C++, semantic networks allow for multiple inheritances. Thus, a particular entity can inherit the features of two or more other categories. In the example in Figure 1.31, *John* is both a boy and a persona (inherits two classes at once) just like *Mercedes* is both a toy and a car.

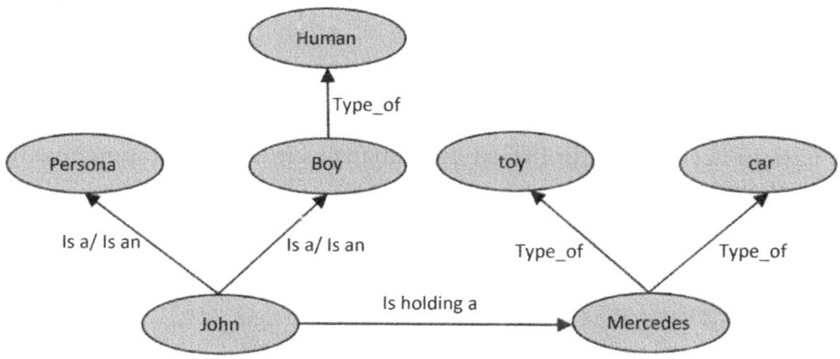

Figure 1.31. *Example of a semantic network with two multiple inheritances*

The problem is that sometimes an entity can inherit conflicting properties for which processing is necessary to obtain a coherent interpretation. For example, the size of a normal car and the size of a toy car are clearly different. Priority must be given to the size of the car in this case for it to be possible for a boy to hold the car in question.

The semantics of semantic networks can be defined in relation to first-order logic, which is their formal equivalent. In their book about artificial intelligence, Russel and Norvig present the rules for converting semantic networks into first-order formulas and vice versa [RUS 95].

Some of the classic limitations often cited about semantic networks concern negation, disjunction and general non-taxonomic knowledge that are phenomena whose processing occurs *ad hoc*.

Semantic networks were the subject of several computer implementations, including the KL-ONE (Knowledge Language One) system. Realized by [BRA 79] (see [WIL 85] for a critical review), this system is distinguished by the use of primitive types (e.g. numbers), so-called generic concepts to express categories as opposed to concepts that express individuals or instances of these categories. Note that WordNet can be considered to be a realization of a semantic network on a large scale to represent the semantic relations between words in a given language [MIL 90].

1.3.1.3. *Conceptual graphs*

Proposed by [SOW 76], conceptual graphs (CG) are a representation formalism based on both the existential graphs of Peirce and semantic networks. Their design was motivated by these objectives:

– the expression of meaning in a precise manner without ambiguities and with an expressive power equivalent to first-order logic;

– the ease of access to information by humans as it is faster to perceive relations between concepts visually;

– the ease of automatic processing by machine, because they have a regular form that simplifies several reasoning, research and indexing algorithms and consequently makes them more efficient.

Formally, a CG is an oriented bipartite graph where the instances of a concept are represented by a rectangle and the conceptual relations are represented by an ellipsis or a circle. The directed arcs connect the concepts and the relations and indicate the direction of the link. Thus, two concepts or relations cannot be directly linked: the connections are always between concepts and relations. When a relation has several arcs, these relations are numbered. For example, the graph provided in Figure 1.32 represents the sentence: *the book is on the table.*

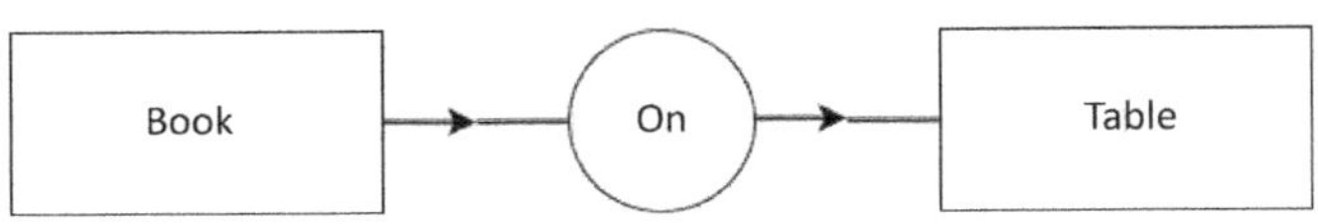

Figure 1.32. *A conceptual graph that represents: the book is on the table*

There is a notation called Linear Form (LF) to represent conceptual graphs in another way than the graphic form, called the display form. LF also serves as a simplified display form. For example, the sentence *the book is on the table* can be presented in the form: [book]-(on)-[table]. Another form of presentation has also been proposed. It is called the *Conceptual Graph Interchange Form (CGIF)*. Our sentence, *the book is on the table*, is represented by the graph: [book: *x] [table: *y] (on ?x ?y). In this graph, *x and *y correspond to a variable definition, while ?x ?y are references to already defined variables. The relation *on* becomes a predicate that connects two arguments: *table* and *book.*

Like with semantic networks, it is possible to translate CGs into an equivalent logical form that is considered to be its semantics. The graph provided in Figure 1.32 is equivalent to the following logical formula: $\exists xy$, book(x) Λ table(y) Λ on(x, y). On the other hand, the graph in Figure 1.33 corresponds to the formula

∀x: book(x) → in_paper(x). The existential quantifier is implied in the conceptual graph. The situation is different for the universal quantifier, which must be indicated explicitly. This means that a sentence like *all of the books are in paper* can be expressed by the graph in Figure 1.33.

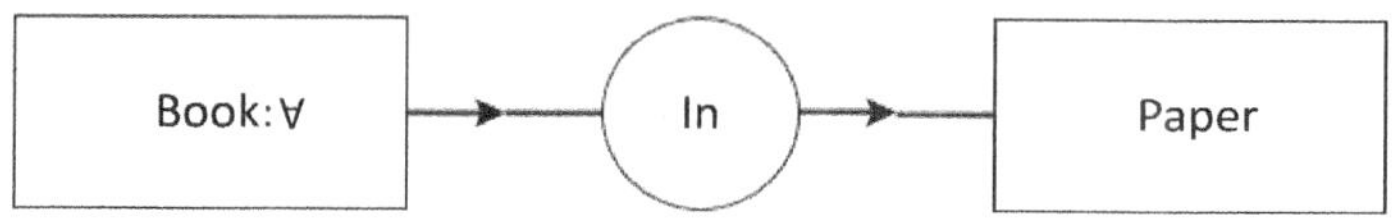

Figure 1.33. *The conceptual graph for the sentence: all books are in paper*

Here again, the advantage of using conceptual graphs compared with the equivalent logical form is practical. Numbers can be expressed in the following form: [cat: @3] *three cats*. It is also possible to express instantiations of objects. For example, we can express the fact that *Matthew is a miner* or that *Lynchburg is a city* in the forms [miner: Matthew] and [City: Lynchburg], respectively.

Conceptual graphs also make it possible to express sentences that connect several entities such as *John goes to Prague by plane tomorrow* (see Figure 1.34).

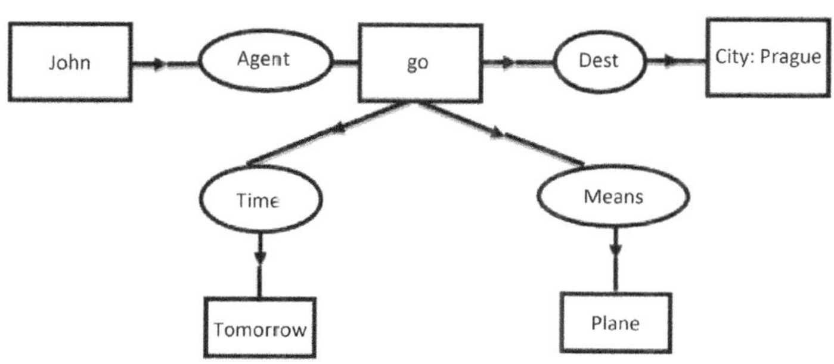

Figure 1.34. *Conceptual graph for John goes to Prague by plane tomorrow*

Represented in the linear form, this gives:

```
[to go]-
  (Agent)    ->    [John]
  (Dest)     ->    [city: Prague]
  (Means)    ->    [plane]
  (Time)     ->    [tomorrow]
```

Sentences with modality (expression of opinion, belief, etc.) can also be translated into conceptual graphs using a nesting process. This is a way to relate the concepts of two or more conceptual graphs. For example, the dependency relations between the components in the sentence *Mary thinks that John wants to go to Prague by plane tomorrow* can be diagrammed in Figure 1.35.

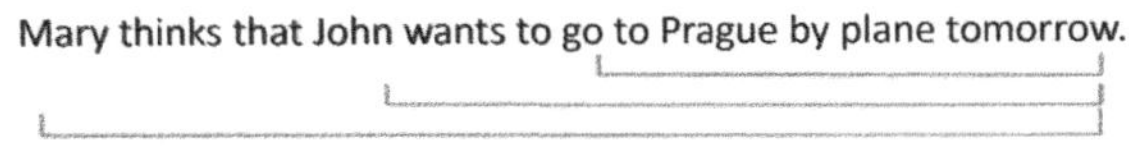

Figure 1.35. *Dependencies between the components of the sentence* Mary thinks that John wants to go to Prague by plane tomorrow

The sentence, whose dependencies are diagrammed in Figure 1.35, can be translated by the conceptual graph presented in Figure 1.36.

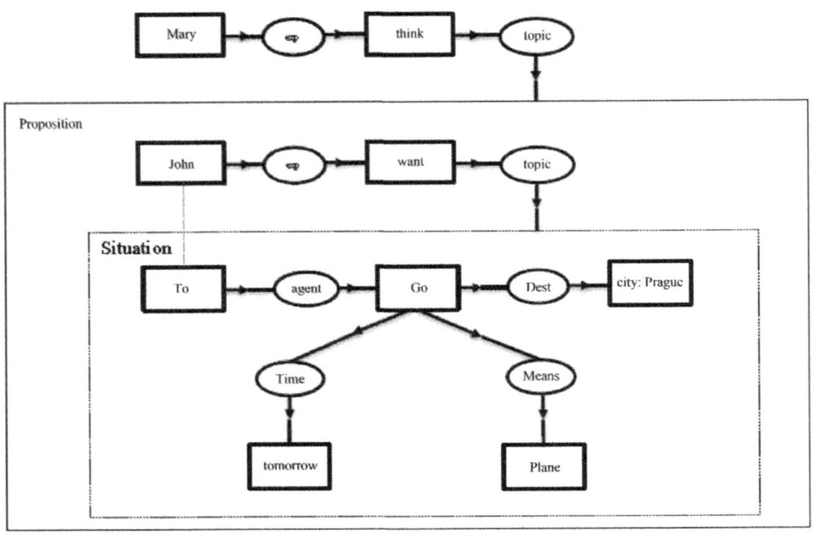

Figure 1.36. *Conceptual graph of the sentence* Mary thinks that John wants to go to Prague by plane tomorrow

In Figure 1.36, a graph can take another graph for an argument. Similarly, it is possible to have particular links between entities such as the dotted line between

John and T, which signifies that this concerns the same person. This graph can have the linear form presented in Figure 1.37.

```
[Mary]<-(Exp.)<-[to think]->(topic)-
        [Proposition: [John *x]<-(Exp.)<-[to want]->(topic)-
        [Situation: [to go]-
          (Agent)    ->    [?x]
          (Dest)     ->    [city: Prague]
          (Means)    ->    [plane]
          (Time)     ->    [Tomorrow]]
```

Figure 1.37. *Linear form of the graph in Figure 1.36*

Conceptual graphs are used in a multitude of NLP applications. Mainly, they have been used to produce a semantic representation of sentences [SOW 86] or texts [ZWE 98]. Conceptual graphs have also been used in information retrieval systems [OUN 98, MON 00]. In these applications, the similarity/distance of the two documents comes down to the similarity of the semantic representations of these documents in the form of conceptual graphs.

1.3.1.4. *Frames*

Initially proposed by Marvin Minsky at MIT at the start of the 1970s, frames are a data structure that store stereotypical knowledge about an object or a concept [MIN 75]. Frames offer a notable advantage compared with semantic networks, which is the presentation of information in the form of feature structures. This allows for a more specific description of entities while representing the relations between the entities.

Within a frame, the information is organized into slots, which are attributes of the entity described by the frame. Typically, a frame includes the following elements:

– The name of the frame.

– The relation between this frame and other frames.

– Slot(s): each slot is a key characteristic of the frame. It can have a digital value (e.g. age: 25, temperature: –7), Boolean (e.g. student or not, military or civilian) or the form of a sequence of characters (e.g. author: Stendhal). Sometimes, a value can be defined by default. For example, a car has four wheels, or a man has two legs.

– Actions associated with the attributes: these actions are generally formulated in the form of if-then rules. For example, in a frame that describes a student, we can formulate the rule: if the grade is <= 50, then the student must retake the exam.

To express generalizations, there are two categories of frames: class frames and instance frames. A class frame can express the properties shared between the members of a class or a given category (e.g. car, book, laborer, etc.). The description of a class is not concerned with an exhaustive description through the enumeration of all features in a class. Instead, it aims to identify the most prominent features that characterize a given class. An instance frame expresses the properties of a particular entity (a particular car, a particular book, a particular laborer, etc.). The relationship between these two types of classes is that of inheritance. Thus, an instance frame inherits the features of a class frame. To clarify this concept, consider the frame presented in Figure 1.38.

Class: laborer

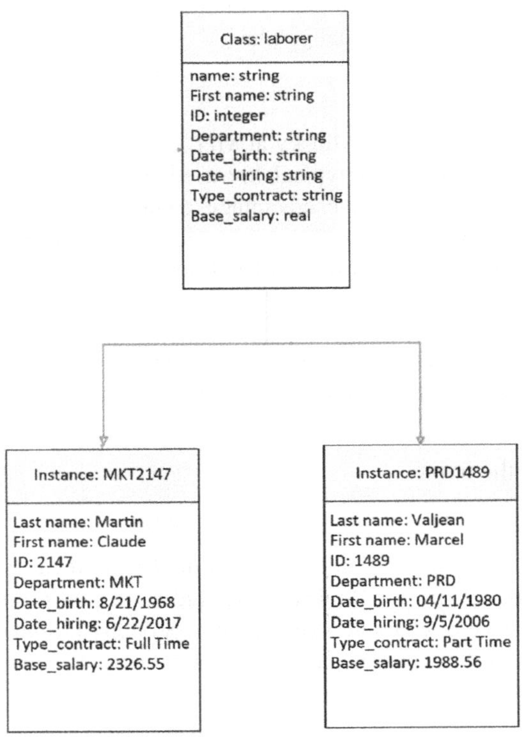

Figure 1.38. *The class laborer and two instances*

It should be noted that the fact that the instances share class properties does not prevent a violation of the values of this class, as the type of contract is an integer in the class, whereas it is a sequence of characters in the (two) instances. Like the object-oriented programming model, frames can be related by three types of relations:

– Generalization: this type concerns relations such as *is_a*, or *sort_of* that connect a *superclass* and a *subclass* where the subclass inherits all of the properties of a superclass. For example, laborer is a sort of employee and, in turn, employee is a sort of person.

– Multiple inheritance is also allowed as a particular entity can belong to different worlds at the same time. For example, Mercedes can inherit from both car and toy.

– Aggregation: concerns the relations of composition or meronymy. Thus, the superclass represents the whole, whereas the subclasses represent the parts. For example, a processor, a hard drive, and a screen are related to a computer by a relation of composition.

– Association: this relation describes the *semantic* relations between classes that are other than those described by the two previous relations. For example, John, Train and Paris are independent classes from the point of view of generalization and aggregation but can be connected by relationships such as *John takes the train to go to Paris*.

To clarify these relations, consider the simplified frame of a plane provided in Table 1.8.

In Table 1.8, the frame of a plane inherits both the properties of a mode of transportation and an instrument of war: a fighter plane can be used to transport its pilots from one airport to another or can be directly involved in an armed conflict. Planes are typically associated with a home airport or a military base. Some of its relations of composition are also represented by the wings and the engine. Note that each of the entities related to a plane also has its own slot. For example, the entity *engine* associated with a plane has its own manufacturer, weight and specific mechanical characteristics.

Despite the advantages related to the simplicity of frames, there are naturally a few disadvantages. In fact, [BRA 85] showed that the authorization of exceptions regarding inherited properties (an entity that shares all but a few of the properties of a superclass) makes it impossible to represent sentences such as “all squares are equilateral rectangles”.

Relation	Type of relation
A plane *is a* machine	inheritance
A plane *is a* mode of transportation	inheritance
A plane *is an* instrument of war	inheritance
A plane *is related to* a home airport	association
A plane *has wings*	aggregation/composition
A plane *has an engine*	aggregation/composition
Role	slot
Weight	slot
Length	slot
Wingspan	slot
Height	slot
Wing surface area	slot
Maximum speed	slot
Rate of climb	slot
Range	slot
Number of engines	slot
Number of pilots	slot
Manufacturer	slot

Table 1.8. *Simplified frame of a plane*

Frames are often used in NLP applications that concern a limited domain. Many task-oriented human–machine dialogue understanding systems have adopted a form of frame for the final semantic representation. The system's task can be summarized by filling in slots in predefined frames that represent elements relevant to the application domain. For example, in the Air Travel Information System (ATIS) domain, the comprehension system seeks to fill a frame with the following information: flight number, city of departure, city of arrival, time of departure, time of arrival, airline, itinerary, etc. (see, for example, [MIN 95, MIN 96, BRA 95]). Applications with richer domains (multi-domains) have also been created but always with frames as the framework for semantic representation. For example, applications in the domains of hotel reservations or tourism information [GAV 00a, KUR 01].

1.3.1.5. *Scripts*

In order to create an automatic comprehension system for English that mimics the cognitive processes of humans, Schank and Abelson at Yale University proposed the concept of a script. It is a means to model conceptual dependencies in order to describe stereotypical event sequences [SCH 77]. This concept offers the advantage of making it possible to predict a particular event in a given context (considering a set of already observed events). It is also an economic means to process information. For example, when someone enters a location like a bank, a train station, bus or restaurant, their actions are strongly predictable. We need only to identify the appropriate script and know the role of the person in this script to find the actions to be done. In new situations, Schank and Abelson consider that humans use plans that underlie scripts. It consists of a repository of general information that makes it possible to connect events that cannot be connected with the scripts available. According to Schank and Abelson, the comprehension process also aims to identify the goal or object of the actors or participants in a story as well as the specific methods they use or that they are prepared to use in order to reach their goal.

The basic elements of a script are:

– Triggering conditions: these are the conditions that must be verified for a script to be triggered.

– The output or the result: this is the final product of the frame's application.

– Properties: the content of a script can be provided in the form of tables or a menu.

– Roles: these are the individual actions of actors or participants. For example, the conductor verifies that passengers have paid for their journey and the driver drives the train, etc.

– Scenes: these are the basic components of a script. For example, the script of the purchase of a train ticket in a train station can be divided into the following scenes: entering the station, locating the appropriate wicket, purchasing the ticket and departing.

Despite often being used as a knowledge base by Natural Language Understanding systems, scripts often suffer from a lack of flexibility.

1.3.1.6. *UNL*

Universal Networking Language (UNL) was designed to code, store and share data independently of language, hence the name "universal". This property makes it particularly well suited for automatic translation, where it represents semantic knowledge. However, UNL, which is similar to semantic networks in several ways, was designed first and foremost for knowledge representation. As it is realistic, UNL does not seek to represent all of the semantic aspects of a sentence but only pertains to the consensual dimensions of them. Thus, the subtleties of poetic language, in the sense of Roman Jakobson, and forms of indirect communication are beyond the objectives intended by UNL (see [CAR 05] for a complete presentation of this language).

Founded in 1996 at the Institute of Advanced Studies (IAS) at the United Nations University in Tokyo, this international project currently includes teams and researchers from all around the world.

Two constraints that are difficult to reconcile govern the writing of UNL expressions. The first constraint is rigor, which means that an expression must provide precise and unambiguous information. The second constraint is that these expressions must be as general as possible to be well understood by the people in charge of developing the translation modules of expressions in human language into UNL (converters) and the translation modules of expressions in UNL into natural languages (deconverters).

Three basic elements constitute the foundation of the syntax in the UNL language: Universal Word (UW) or virtual vocabulary items, relation labels and attribute labels. UWs are words that transmit knowledge or concepts. They correspond to the nodes in a UNL graph. Two types of UWs can be distinguished: permanent UWs and temporary UWs. Permanent UWs correspond to concepts of common use and are included in the UW dictionary. Temporary UWs correspond to new concepts that are specific or difficult to translate. English was adopted to establish the UNL vocabulary because this language was the most well known by the majority of researchers involved in the project. Semantic relation labels decorate the arcs that connect the nodes of a UNL graph (UWs). Attribute labels are information such as number, gender, aspect, mood or emphasis. They are expressed by features independent of language. Consider the sentence [1.5], which can be presented by the UNL graph in Figure 1.39.

Mary hit John with a stick at the cinema yesterday because of Nicole. [1.5]

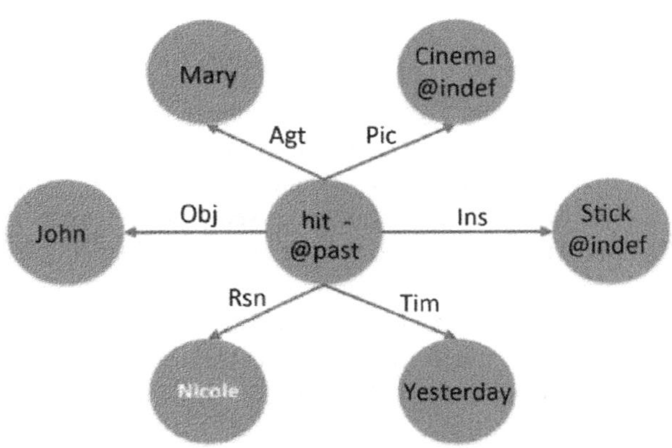

Figure 1.39. *UNL graph of sentence 1.6*

In the UNL graph of sentence [1.5], the three levels of representation are:

– UWs: Nicole, yesterday, stick, John, hit, Mary, cinema.

– Relations: agt (agent), obj (patient), tim (time), ins (instrument), plc (place) and rsn (reason).

– Universal attributes: @past (past), @def (definite) and @indef (indefinite).

The formal difference between UNL and semantic networks resides in the linguistics constraints imposed. A relation can be of any kind in a semantic network – linguistic, biological or physical – whereas the number of relations is fixed in the framework of UNL formalism. This formalism is at the core of an environment including a product, tools, data, a community, etc.

1.3.2. *Ontologies*

The word *ontology* can be analyzed as two morphemes: *on* or *ontos*, which signifies being, and *logos*, which signifies science or discourse. This philosophical term from the 17th Century, often written with a capital *O*, concerns the part of metaphysics that pertains to being in its essence, independent of the phenomena of its existence [ENC 09]. Ontology with a capital *O* attempts to address questions like: *what is a being?* Or: *what are the features shared by all beings?* Another philosophical use of the term, with a lowercase *o*, signifies a categorization system that accounts for different perspectives about the world. For example, depending on the person, reptiles can be seen as repulsive or frightening animals, pets or a promising research subject.

The modern understanding of ontology is located at the intersection of philosophy, artificial intelligence and lexical semantics. It designates a particular vision of a specific domain that is shared by a group of people and is used as a framework in the goal of resolving a particular problem [USC 96]. The main difference between an ontology and knowledge is that an ontology is independent of language, it is generic, it can be enriched and it is available in a digital format that is easy to manipulate with a computer (see [GUA 95] for a more detailed discussion).

Three main reasons are often cited to justify the use of ontologies in computer systems [USC 96]. First of all, they are a good way to disambiguate key concepts in any domain. They facilitate the emergence of a common, or at least similar, understanding of the problem by members of a team that all have different points of view depending on their disciplines and work context. Second, it is also often necessary to share the same information between different modules in the same system, which concerns interoperability. To do this, an ontology serves as an intermediary between these different modules, whose functions and algorithms can vary considerably. Finally, in the particular case of a computer system with an NLP module, the use of an ontology makes this system more generic and less dependent on language. It is therefore necessary that the output of the NLP module is compatible with the ontology, in order to facilitate access to information (see [MAS 07] for a discussion of the use of a high-level ontology in the framework of a multi-agent system).

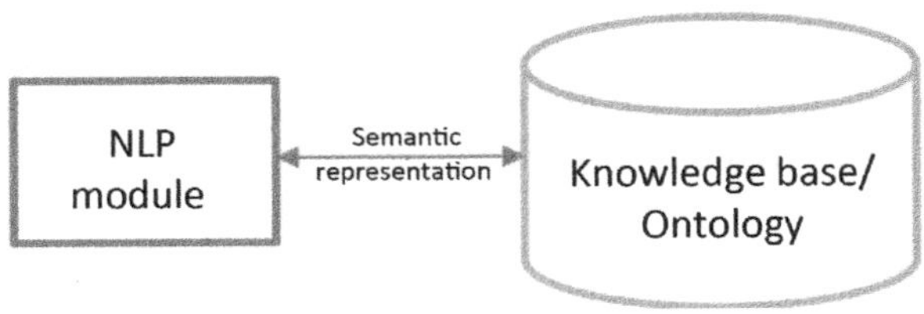

Figure 1.40. *Global architecture of an information system with an ontology and an NLP module*

In Figure 1.40, the semantic representation produced by the NLP module depends directly on the ontology. The information that this module produces, the level of finesse of the representation and the form of the representation (logical representation, frame, etc.) depend directly on the choices made in the ontology. Consider the word *mouse* as an example. Evidently, this word is ambiguous because it has several meanings, including: a small rodent mammal and a manual computer input device whose movement shifts the cursor. In the case of an exchange between two agents who share an ontology about material computer equipment, such an ontology would include a description of a mouse as a part of the computer. In this case, the ontology plays the role of foundational knowledge that serves to rectify incorrect references and consequently allows for disambiguation (see Figure 1.41).

Certain tasks, such as anaphora resolution, sometimes require extra-linguistic knowledge that can only be found in an ontology. For example, to produce a semantic representation of sentence [1.6], an inference must be made about the domain to know what object has an engine, can be driven. An ontology can offer answers to these sorts of questions.

He drove it too fast, which is why the engine broke down. [1.6]

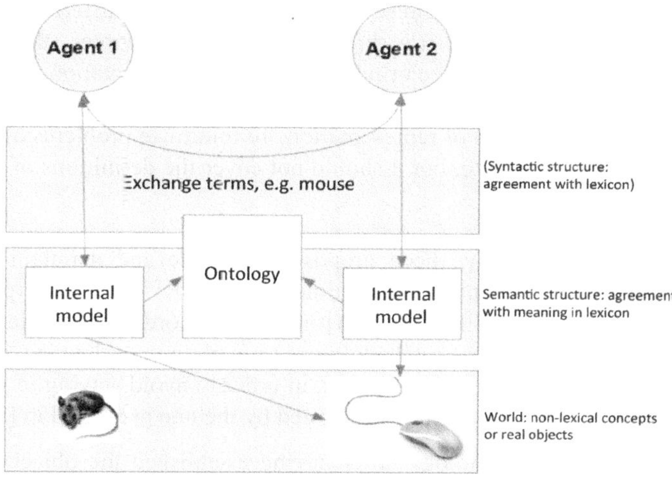

Figure 1.41. *Role of an ontology in an information exchange process [MAE 03]*

1.3.2.1. *Methodology for developing ontologies*

Before presenting the methodologies used to develop ontologies, it seems pertinent to address the question of the criteria of a good ontology. [GRU 95] proposed the following criteria[15]:

– Clarity: ontology must allow for communicating the meaning of undefined terms. These definitions must be objective and independent of the social or computing context. They must also be documented in natural language.

– Coherence: inferences made based on the ontology must be in agreement with the definitions of concepts that it offers.

– Possibility of extension: the ontology must be designed with consideration for the possibility of future extensions. This can concern the addition of new usages, whether they are specializations or generalizations. Similarly, it must be possible to add the definitions of new terms without needing to modify the definitions of existing terms.

– Minimal encoding deformation: as much as possible, deformations in the conceptualization that result from the specification should be avoided.

15 Gruber states that these criteria are particularly relevant to ontologies intended for information sharing, but they also apply to other types of ontologies, albeit to a lesser extent.

– Minimal encoding bias: because it is possible that the different modules and agents that share an ontology use different knowledge sources, there are cases where, for reasons of convenience or otherwise, some representations are adapted to the notation or implementation system, which creates a bias when carrying over this knowledge to another system of representation. To minimize problems of portability, an ontology must be complete, but it should not cover the definitions of superfluous terms.

Many methodologies have been proposed to create and maintain ontologies, including the TOVE (Toronto Virtual Enterprise) [GRÜ 95], Enterprise Model Approach [USC 95] and IDEF5 [PER 94] methods. In order to be brief, I will not include a complete list of these methodologies or explain their details (see [JON 98] for a more detailed introduction). However, in order to avoid staying in generalities, here is an abridged version of the steps inspired by the one presented in [USC 96]:

– Construction of the ontology: after having established the objectives and the range of the ontology, the key concepts of this ontology must be identified and listed. Defining each of the concepts identified in natural language makes it possible to minimize ambiguities and facilitate communication within the work team. Then, the ontology is coded using a language that is deemed appropriate. In some cases, the construction of the ontology does not start from zero. In that case, this concerns extending or adapting an existing ontology. This requires a very detailed study of the concepts of the existing ontology in order to avoid redundancies and to correctly process similar concepts.

– Evaluation: this is a qualitative technical judgment regarding the adaptation of the ontology in relation to its reference framework [GÓM 95]. The reference framework is the system prerequisite in the usage environment.

– Documentation: the effective use of an ontology requires clear documentation. The documentation must include an explanation of the main assumptions about the concepts, as well as the primitives used to explain these concepts. Note that some existing tools offer publishing environments and/or semi-automatic aids to write the documentation.

1.3.2.2. *Structure of an ontology*

According to [BAC 00], ontologies are intimately connected with a formal language. He proposes an alternative to the definition provided in the previous section that he considers both precise and rigorous: "to define an ontology for the representation of knowledge, is to define, for a given domain and problem, the functional and relational signature of a formal language of representation and the associated semantics". For his part, [GUA 98] proposes a logical vision of an ontology that he considers to be a system capable of accounting for the intentional meaning of a formal vocabulary. According to him, the main difference between ontology and conceptualization resides in the fact that ontologies are dependent on language while conceptualization is independent of it.

An ontology is a set of concepts in a given domain and the relations between those concepts. It is used to think about the properties of this domain and sometimes to define the domain. Unlike knowledge representation, an ontology has a more generic level of description.

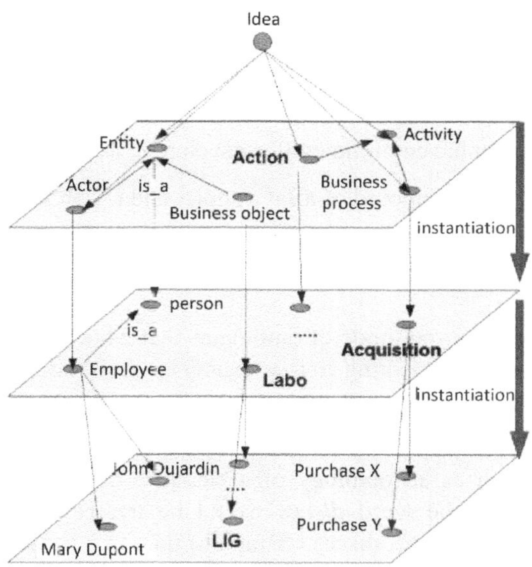

Figure 1.42. *The levels of knowledge in an ontology (adapted from [RIG 99])*

Three levels of knowledge can be distinguished within an ontology: methodological knowledge, conceptual knowledge and factual knowledge (see Figure 1.42).

– Methodological knowledge consists of a high-level language to express ideas and meta-types. Several resource organization languages on the Internet are used as a framework to express methodological knowledge such as the XML and RDF (Resource Description Framework) language. Logical descriptive languages, such as OWL (Web Ontology Language), are also used for this objective.

– Conceptual knowledge is necessary for understanding the meaning of objects like meronymic knowledge: a plane is composed of wings, ailerons, jet engines, a cockpit, etc.

– Factual knowledge concerns information about objects in the real world, for example, the LIG laboratory in Grenoble has 24 research teams, the Harvard University endowment fund was worth approximately 30 billion dollars in 2013, etc.

1.3.2.3. *Tools for developing ontologies*

To develop an ontology, there are many tools available, including:

– Protégé[16]: a free tool for publishing ontologies and knowledge bases (probably the most well known).

– Chimaera[17]: a system to create and maintain distributed ontologies. It can also troubleshoot an ontology or merge several ontologies.

– OntoEdit[18]: a graphic environment for developing and maintaining ontologies.

– WebOnto[19]: a tool composed of a Java applet and a web server that can publish ontologies and navigate in them.

1.3.2.4. *A few ontologies*

There are currently a multitude of ontologies that have very different properties and objectives. Before presenting a few examples, it is necessary to distinguish between the two main types of ontologies:

– Domain ontologies represent the particular meaning of terms as they apply to a specific domain, such as an ontology of agriculture or an ontology of computer science. For example, the word *discus* would be treated differently in a sports ontology (throwing the discus, discus champion) than in a pet ontology (a beautiful Amazonian fish).

– Foundation ontologies concern the modeling of common objects that are usable through a fairly vast set of domain ontologies. They include a glossary in which the terms can be used to describe a set of domains. Sometimes called top-level ontologies, this type of ontology is mainly used for the semantic integration of several domain ontologies as well as the development of new ontologies.

In the literature, there is a rather large number of ontologies of various types including the General Formal Ontology (GFO), Unified Foundational Ontology

16 http://protege.stanford.edu/
17 http://www.ksl.stanford.edu/software/chimaera/
18 http://www.daml.org/tools/#OntoEdit
19 http://kmi.open.ac.uk/technologies/name/webonto

(UFO), Business Object Reference Ontology, Cyc, etc. This text includes three that were selected as representative.

Developed by Nicola Guarino and his collaborators at the Laboratory for Applied Ontology at the Italian NRC, the Descriptive Ontology for Linguistic and Cognitive Engineering (DOLCE)[20] is an ontology that does not have a universalist vocation [MAS 03]. Instead, it is a point of departure to clarify the assumptions behind existing ontologies or linguistic resources like WordNet [GAN 03]. As its name indicates, DOLCE is an ontology that was designed to reflect language and human cognition. It is based on the KIF language and contains about 100 concepts and a similar number of axioms.

As shown in Figure 1.43, DOLCE has many fundamental distinctions. The primary focus here is on the distinction between the Endurant and Perdurant entities, qualities and qualia.

Endurant entities are wholly present at all moments of their existence. For example, a laptop, a table or a dress are entities that exist in time. Perdurant entities have a partial existence at a given moment in their existence. These are entities in the course of being carried out. Perdurant entities correspond to processes like reading, a kick, or rain, or other processes that are only partially present at a given moment in its existence.

In addition, the distinction between qualities and qualia merits examination. Qualities correspond to the properties of an entity like color or temperature, whereas qualia are the perceptive representations of qualities. Each type of quality has its own qualitative space.

An extension of DOLCE was developed by Aldo Gangemi at the STLab in Rome. It is DnS[21] (Descriptions and Situations). It is distinguished by the fact that it does not put restrictions on the types of entities and relations that can be applied in the context of a domain specification and can be seen as a top-level ontology.

Freely available, SUMO[22] is a formal ontology. Its 2002 version contained about 1,000 terms and 3,700 definitions [PEA 02]. It is coded based on a first-order logical format called Standard Upper Ontology Knowledge Interchange Format (SUO-KIF), which is a simplified form of the KIF language [PEA 09]. This ontology covers domains like the temporal and spatial domains and is the object of extensions in domains as varied as finance and terrain and weather modeling. The structural

20 Descriptive Ontology for Linguistic and Cognitive Engineering.

21 http://www.loa.istc.cnr.it/ontologies/ExtendedDnS.owl

22 Suggested Upper Merged Ontology.

ontology consists of a set of definitions of some syntactic abbreviations on the basis of vocabulary provided by SUO-KIF. The base ontology is comprised of a top-level concept hierarchy that includes the sections: set/class theory, numeric, temporal, mereotopology (see Figure 1.44).

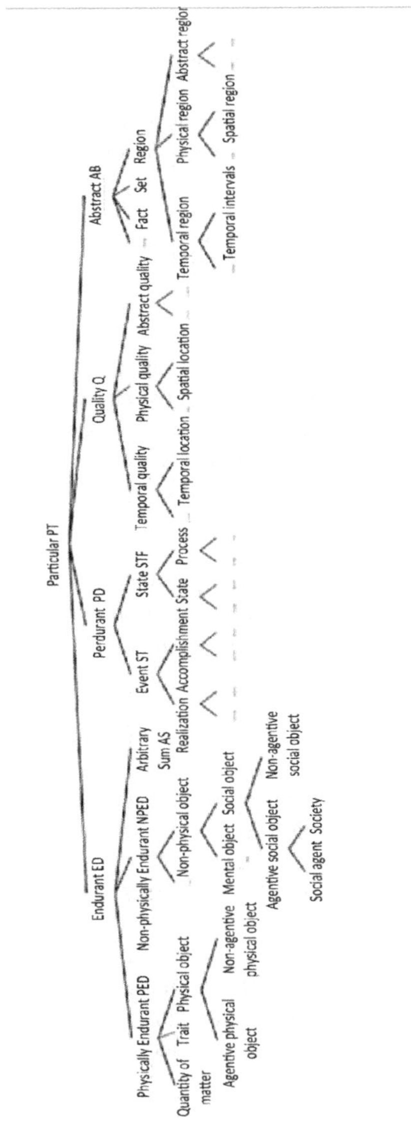

Figure 1.43. *Taxonomy of DOLCE [MAS 03]*

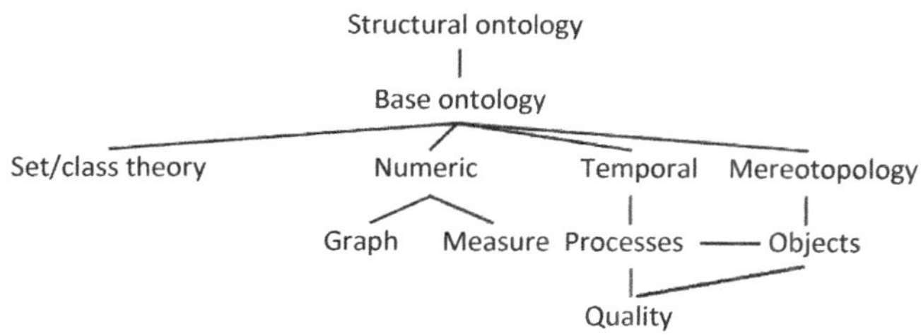

Figure 1.44. *Base structure of the SUMO ontology*

A series of top-level distinctions operate within SUMO. Abstract entities are distinguished from physical entities. In turn, physical entities are divided into two groups: objects and processes. Several types of processes are also distinguished: binary processes that affect two objects, internal changes, biological changes, chemical processes, creation, etc.

A mid-level ontology has been added to SUMO. It includes elements like: communications, countries and regions, airports in the world, and viruses, with the goal of connecting several domain ontologies to SUMO.

As a complete initiation to the KIF-SUMO language is beyond the scope of this presentation, only a few examples are provided in order to give a more concrete idea of this language (see Figure 1.45). For a more detailed representation of this language, see [PEA 09].

Finally, it should be noted that SUMO has templates and a lexicon in English, German, Czech, Hindi and Chinese to allow multilingual generation. A link between SUMO and WordNet has also been established [NIL 03].

The Basic Formal Ontology (BFO)[23] was initially proposed by Barry Smith and his collaborators. It is a formal ontological framework that consists of a series of sub-ontologies at different levels of granularity [SMI 04]. The BFO was developed in the context of the *Forms of Life* project funded by the Volkswagen foundation. BFO has been extended for applications in several domains including bioinformatics [GRE 04]. It consists of a set of sub-ontologies that have various levels of granularity.

23 http://ifomis.uni-saarland.de/bfo/

(subclass Human Mammal)	The class *human* is a subclass of the class *mammal.*
(and (instance FrançoisHollande Human) (occupiesPosition FrançoisHollande PresidentFrance))	François Hollande is a human. He occupies the position of President of France.
(or (instance BillGates Human) (occupiesPosition BillGates CEO Microsoft))	The operator *Or* is not equivalent to *or* in natural language. It signifies two things: one or the other or both things at once.
(not (occupiesPosition BarackObama MayorDenver))	Barack Obama is not the mayor of Denver

(=>
 (and
  (instance ?H Human)
  (attribute ?SL to sleep))
 (not
  (exists ?ACT
   (and
    (instance ?ACT ProcessesIntentional)
    (overlaps ?ACT ?SL)
    (agent ?ACT ?H)))))
	A person cannot commit an intentional act when they are in the process of sleeping.

(forall (?C ?T)
 (=>
  (and
   (instance ?C child)
   (instance ?T toy))
  (likes ?C ?T)))
	All children like toys.

Figure 1.45. *Examples of representation with the language KIF-SUMO*

A top-level ontology like BFO is based on the distinction between basic entities, hence the necessity of defining the notion of a boundary between the entities [SMI 97, VOG 12]. Two types of boundaries have been identified in the literature: external boundaries and internal boundaries. To explain these two types of boundaries, consider the following entities: Mary, Neptune and an eraser. Each of these objects has an external boundary. They include Mary's skin, Neptune's surface and the eraser's surface, respectively. The internal boundaries are boundaries that separate the parts of an entity from one another. For Mary, this would be her organs (heart, lungs, eyes, etc.) or cells. For Neptune, this would be its layers: surface layer (which has particular properties), core, etc. On the other hand, in the case of the eraser, it is a bit different because the eraser is a materially homogenous object. In this kind of case, the boundaries are rather functional or conceptual. For example, this would distinguish the surface of the eraser that is directly eroded by rubbing it on paper.

Thus, there are two types of internal boundaries. The first type includes boundaries that assume a certain material discontinuity, for example, because of the existence of holes, fissures or tears. These are called Bona Fide (BF) Boundaries[24]. There are also external boundaries of this type. For example, in the real world, we often talk about the boundaries between properties (land, farms, etc.); some bodies of water like the Mediterranean Sea and its *vague* boundaries with the Adriatic, Ligurian, Tyrrhenian and Alboran seas; administrative regions like *Texas*, *Virginia* and *California*; counties like the *Orange County*, *Napa County* and *Riverside County*; or countries like France, Gabon and Syria. These boundaries, although essentially the result of a *social* agreement of a larger or smaller community of people, involve very real rights and obligations. A violation of one of these boundaries can result in legal, administrative or military conflicts. Other BF boundaries can have a mathematical dimension like the equator, the Tropic of Cancer, etc., or an individual dimension like a geographic zone to cover some kind of work. In other words, BF boundaries are artificial boundaries that depend on human perception, which is subjective by nature.

Fiat Boundaries (FB) are boundaries that involve a material heterogeneity, such as material construction, texture or electric charge, which separates them from their surroundings. Consequently, they do not depend on the mind of a given subject and are therefore natural. By extension, objects that have BF boundaries are called BF objects and objects that have Fiat boundaries are called FO objects.

The BFO model proposes a distinction between two types of entities: SNAP entities and SPAN entities [SMI 04].

SNAP entities are characterized by a continuity over time and the preservation of their identity. They have a total presence at all moments of their existence (see Figure 1.46). For example, the specific yellow of a particular pear. Three types of these ontologies can be distinguished:

– Independent entities: the substances and parts whose existence does not depend on another entity. For example, apple and car are independent entities.

– Dependent entities: concern qualities, roles (like professor, taxi driver or soldier), conditions, functions (like the function of a pen that permits writing),

24 The two types are: *Bona Fide Boundaries* and *Fiat Boundaries*.

power (of an engine, for example). The role of professor cannot exist independently of the person who possesses this quality.

– Spatial regions: whether they are in zero, one, two or three dimensions.

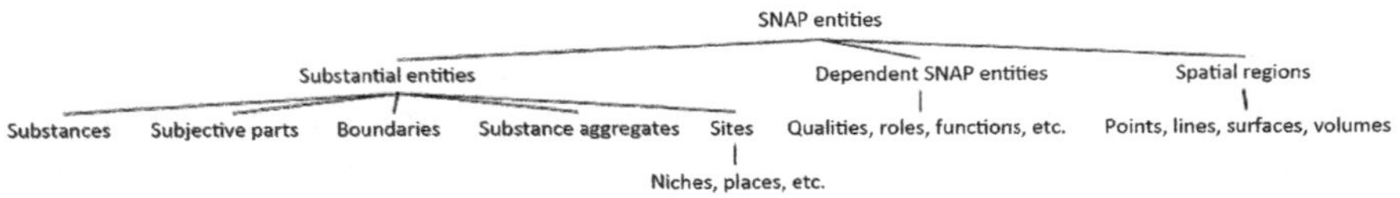

Figure 1.46. *Hierarchy of SNAP entities*

On the other hand, SPAN entities have temporal parts and are deployed phase by phase and exist only in their successive phases (see Figure 1.47).

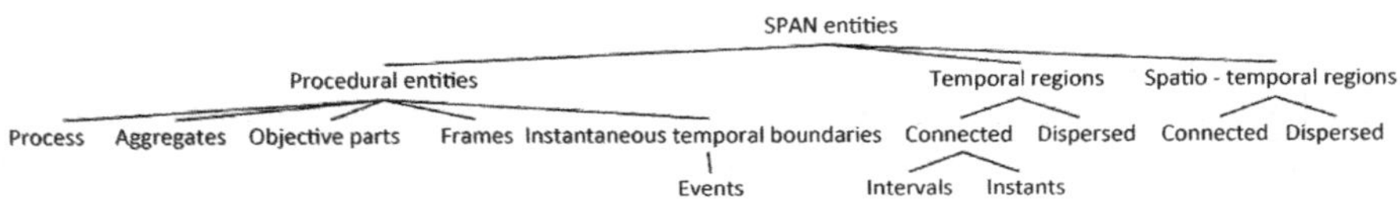

Figure 1.47. *Hierarchy of SPAN entities*

For example, the period of a person's youth, a period of study and wars are SPAN entities.

2

The Sphere of Semantics

2.1. Combinatorial semantics

Far from the theoretical or psycholinguistic controversies about the role and manner in which humans combine different resources of linguistic information to construct meaning in a sentence, it is generally accepted that the lexicon and syntax play a non-negligible role in this process[1]. Because these two sources of information are not always sufficient, high-level semantic constraints are indispensable for distinguishing different possible interpretations of the same syntactic structure. For example, the subject and the object of a sentence can play different semantic roles. Sometimes, the subject is the direct cause of an event that is voluntary or not (e.g. John writes), or the indirect or unconscious cause (e.g. it is raining). Sometimes, it is the subject that is affected by the action expressed by the verb (as in, John dies).

2.1.1. *Interpretive semantics*

At the start of the 1960s, Fodor and Katz [FOD 64] proposed enriching syntactic structures constructed in the form of trees with a semantic analysis.

1 The way in which syntactic and semantic knowledge intervenes in the processing of a sentence is the subject of significant controversy in the domain of psycholinguistics. On the one hand, some like [TYL 77, TRU 92] insist that semantics intervenes before the end of the utterance analysis, whereas others believe that syntactic analysis occurs before, independent of semantics [RAY 83, RAY 92]. See [CRO 96, MAH 95] for a general overview of these works.

Interpretive semantics establishes a clear distinction between aspects of interpretation that are founded on linguistic knowledge and aspects of interpretation that are derived from knowledge about the world. According to this theory, it is this distinction that makes it possible to draw the boundary between semantics and pragmatics. It assumes that semantics must account for all of the possible interpretations of a given sentence independent of the limits imposed by knowledge about the world. The objectives of this theory can be summarized by the following points:

– to determine the number and content of possible interpretations of a sentence;

– to detect semantic anomalies;

– to decide whether there are paraphrase relations between two sentences;

– to indicate all other semantic properties that play a role in this capacity.

According to Katz and Fodor, semantics is based on two fundamental components: a dictionary and projection rules. In this context, it is assumed that grammatical analysis is pre-existing.

The role of the dictionary is to provide information about the parts of speech (noun, verb, adjective, etc.) as well as a description of the role played by the words in the semantic interpretation in the form of *marker* or *features* such as: (±human), (±concrete) and (±animated). The lexicon also contains so-called lexical redundancy rules that represent the relations between these features. For example, the rule (+human) → (+animated) indicates that everything that belongs to the class of humans also belongs to the class of animated beings.

For example, the word *pig* can have the description given in Figure 2.1.

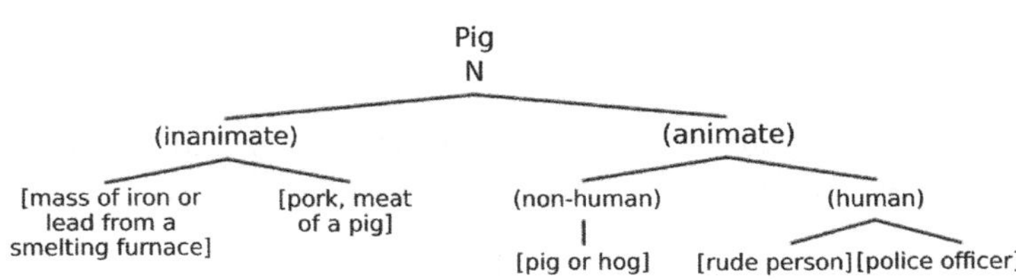

Figure 2.1. *Description of the word pig according to the model in [FOD 64]*

In Figure 2.1, the syntactic categories are represented without parentheses, the semantic categories are in parentheses, and the semantic features are in square brackets.

The projection rules define the way in which the different aspects of the meaning of a lexical item can be combined in the framework of grammatical constituents. In doing so, they indicate how the set of sentences encountered by a locutor are projected on the infinite set of grammatical sentences in language.

To illustrate these ideas, consider the following sentence as an example: the man drives the colorful car[2]. In this seemingly simple sentence, there are three cases of lexical ambiguity. The first concerns the word *to drive* which can be interpreted in at least four ways, including[3]: to operate a vehicle, to cause someone or something to move, to travel, and to motivate, compel or force. The second word, *car*, can in turn have at least three different meanings: automobile, conveyance for passengers on a cable railway, and passenger compartment of an elevator. Another ambiguity concerns the word *colorful*, which can have at least two different interpretations: that which has one or several colors, or that which is vivid or lively. In theory, the number of possible interpretations of a sentence is equal to multiplying the number of interpretations for each ambiguous word. In this case, this means there are 24 interpretations: 4*3*2. Projection rules intervene here and exclude some combinations through selective restrictions that specify the necessary and sufficient conditions. For example, the word *to drive* in the sense *to motivate* has a selective restriction that limits its occurrence to animated objects (to motivate your friend, to motivate your fellow citizens, etc.) and it therefore cannot be employed with an inanimate object like *car*. Similarly, the adjective *colorful* in its figurative sense meaning *vivid* and *lively* requires a noun that designates an abstract entity (e.g. ideas, style, expression) and cannot qualify a noun that designates a concrete entity like *car*.

The work by [KAT 64] concluded that all of the information necessary for applying projection rules can be found in the deep structure. In other words, the transformation operations do not affect the meaning of the final sentence. Known today as the Katz–Postal hypothesis, this conclusion has been supported by several observations. First of all, the rules of the passive modify the grammatical relations of the sentence. It seemed logical to apply projection rules to a structural level that pre-exists the application of these rules: deep structure. Second, discontinuity phenomena were typically created by transformation rules whereas discontinuous deep structures that became continuous after transformations had never been observed. This was seen as an argument in favor of their interpretation at the deep level where the semantic unit is reflected by its syntactic continuity.

2 This example is inspired by the example originally provided by Katz and Fodor themselves: *The man hit the colorful ball.*

3 For simplicity's sake, only the meanings deemed most prominent for each of the words have been provided.

This perspective of the role of semantics is consistent with the standard theory by [CHO 57, CHO 65]. This theory stipulates the existence of formal rules for which the role is to generate sentence skeletons that are then lexically enriched by lexical insertion rules to create the deep structure. Finally, transformation rules are applied to the deep structure to produce surface structures such as negative, interrogative or passive structures. Thus, the role of semantics is to interpret the deep structure (hence the name interpretive semantics) which, after the transformations, receives a phonetic interpretation.

Later, Chomsky and his students began to recognize that certain properties in the surface structure like accentuation played a role in the semantic interpretation of the sentence, as in sentences [2.1] where the accent is marked by an apostrophe:

– Mary bought her laptop at the store'. [2.1]

– Mary bought her laptop' at the store.

Although these two sentences are syntactically identical, they are not semantically equivalent: in the first sentence, the information is shared between the locutor and the recipient, whereas in the second sentence, only the locutor has the information in advance. Thus, the location of the accent, a surface pronunciation phenomenon, contributes to a change in the focus of the sentence and therefore a semantic change.

Using examples like [2.2a] and [2.2b], Jackendoff demonstrated that passive transformation plays a role in the interpretation of the sentence:

a) Many arrows did not hit the target. [2.2]

b) The target was not hit by many arrows.

The scope of many in [2.2] seems larger in sentence a than in sentence b. The word order, which is also a surface phenomenon, seems to play a semantic role. Consider sentences [2.3].

a) John <u>only</u> reads books about politics. [2.3]

b) <u>Only</u> John reads books about politics.

The scope of the quantifier *only* is not the same in sentences [2.3a] and [2.3b]. In the first case, it pertains to the object complement and in the second case, it concerns the subjects.

Most researchers from the interpretive school agreed with Chomsky that deep structure alone is not sufficient to provide all of the elements necessary for the interpretation of the sentence [CHO 71]. This new view of syntax and semantics is generally called Extended Standard Theory.

During the 1970s, Ray Jackendoff proposed a more complete model of interpretive semantics [JAC 72]. It consisted of a distributed model where there was no single semantic representation, but rather several types of rules that apply at several levels. Four components of meaning, each of which is derived from a group of interpretive rules, are distinguished: the functional structure, the modal structure, the table of coreference, and focus and presupposition.

The functional structure specifies the main propositional content of the sentence. It is determined by projection rules that apply to the deep structure. For example, to resolve the problem of the passive transformation shown in sentences [2.2], there are rules such as:

– the subject of the deep structure of the phrase must be interpreted as the semantic agent of the verb;

– the direct object of the deep structure must be interpreted as the semantic patient of the verb.

The modal structure specifies the scope of logical elements such as negation and quantifiers as well as referential properties like nominal groups. The following rule is given as an example: if logical element A precedes logical element B in the surface structure, then A must be interpreted as having a wider scope than B where the logical elements include quantifiers, negatives and some modal auxiliaries.

The table of coreference concerns the way in which elements like pronouns are considered to be coreferential with their antecedents. This was contrary to the dominant idea at the time, according to which sentences with anaphors were the result of a pronominalization transformation. Thus, sentence [2.4] is the result of the transformation:

a) Frank thinks that Frank will succeed this year. [2.4]

b) Franki thinks that he will succeed this year.

Thus, sentence [2.4a] is considered to be a transformation of [2.4b]. This explanation, although it is appealing, does not account for sentences with crossed coreferences such as the ones discovered by Emmon Bach [BAC 89]:

– [The man who deserves itj]i will get [the prize he desires]j [2.5]

The existence of a reference from one nominal group to another, which in turn refers to the first one, involves an impossible deep structure: every anaphor can be found in the antecedent of another.

Focus and presupposition concern the distinction between the parts of the sentence that are considered to be new (providing new information) or those that are old.

Toward the end of the 1970s, ideas about deep structure and transformation began to give way to more functional approaches like in [BRE 82].

2.1.2. *Generative semantics*

Developed in the mid-1960s in response to the interpretive semantics of Fodor and Katz, generative semantics stipulates that the semantic component is generative while the syntactic component is interpretive. In other words, according to this approach, the role of the syntactic component is to provide appropriate structures to the meaning of the sentence that plays the role of deep structure through transformation operations (see the diagram in Figure 2.2). This approach is attributed jointly to several researchers including George Lakoff, James McCawley, Paul Postal and John R. Ross.

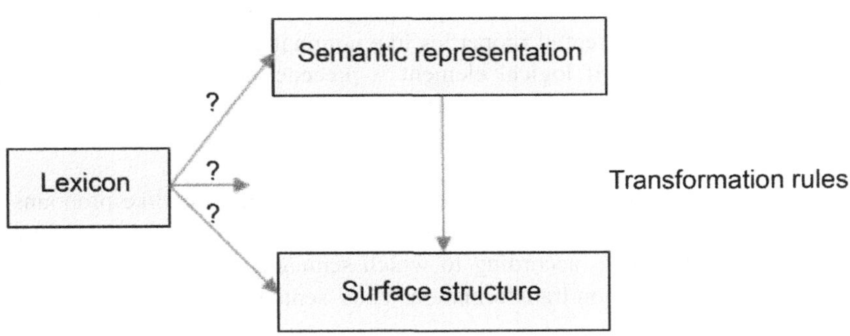

Figure 2.2. *General diagram of the initial model of generative semantics*

This amalgamation of syntax and semantics led to the formulation of three main hypotheses that are inherent in generative semantics. The first stipulates that deep structure as conceived by Chomsky in his book *Aspects of the Theory of Syntax* [CHO 65] does not exist. According to the second hypothesis, the initial representations of the derivations are logical. They are thus independent of language based on the hypothesis of a universal base. Finally, the derivation of a sentence is a connection between a semantic representation and a surface form.

Note that in Figure 2.3, the place and the way in which the lexicon intervenes in the derivation is controversial. Among others, [MCC 76] proposed a solution that treats lexical entries as semantic structures. For example, according to this approach, the verb *to kill* is derived from an abstract semantic representation: to cause to become not alive. This representation gives the following structure to the sentence *John killed Mary*. This deep structure is expressed semantically in terms of metalinguistics such as to cause, alive, etc.

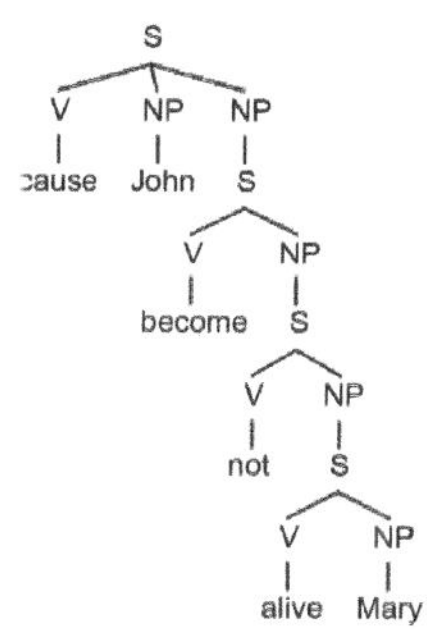

Figure 2.3. *Semantic representation of the deep structure for: John killed Mary*

To move from the deep structure to the surface structure, several predicate raising operations are carried out on the semantic tree. These operations make it possible to raise the verb from one embedded sentence to the next higher level, where it is attached to the verb in the sentence at that level. Thus, *alive* is raised to a higher level where it is attached to *not*. The result of this operation is given in Figure 2.4.

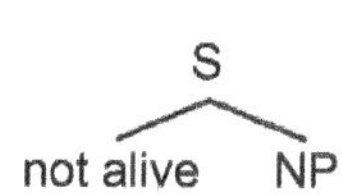

Figure 2.4. *Result of a predicate raising operation*

The successive application of such operations leads to a combination of the constituents *to cause, to become, not* and *alive*. Finally, generativists and interpretivists did not consider their respective theories as being cognitive models of the production or analysis of sentences. This led some linguists to conclude that the debate between the two camps had no point.

2.1.3. *Case grammar*

Proposed by Charles Fillmore at the end of the 1960s as a major modification to transformational grammar, case grammar quickly gained popularity, especially in the United States [FIL 68]. As in many other linguistic theories, there is no unified form of case grammar. In fact, Fillmore proposed several successive versions in the 1960s and 1970s [FIL 66, FIL 71, FIL 77].

Unlike classic cases observed in languages like Latin, Russian and Arabic, where the functions and/or semantic roles are signaled using suffixes, the cases proposed by Fillmore are more general and serve to model the relation between the verb, which plays the role of a logical predicate, and nominal groups in the sentence, which are treated like participants[4]. Two important properties are attributed to cases: they are universal and only a limited number exist.

The semantic representation obtained in this approach is considered to be the deep structure of the sentence. Thus, it is necessary to separate the grammatical functions (subject, object, etc.) and the cases that represent the underlying semantic relations between the participants in the situation evoked by the verbal predicate. For example, in sentences [2.6], John is considered to be an agent, although he has the same syntactic function in the surface forms of the sentence:

(John subject/agent) hit Peter. [2.6]

(**John** subject/recipient) hit the jackpot.

In addition, the same semantic role can be realized by different grammatical functions, as in sentences [2.7]:

The sun is drying (the wheat object/patient). [2.7]

(The wheat subject/patient) is drying.

In his founding article, [FIL 68] proposed the six following cases (this list is considered non-exhaustive)[5]:

– Agent: the case of an animated entity perceived to be the instigator of the action expressed by the verb.

– Instrument: the case of an inanimate force or object involved in the action or state expressed by the verb.

4 Note that, as emphasized by Fillmore, at that time, linguists in the generativist movement like Lyons considered cases (in languages where they exist) to be inflectional realizations of certain syntactic relations on the surface level.

5 Other lists were proposed by Fillmore later on as well as by other researchers (see [AND 85] for another example).

– Dative: the case of an animate being affected by the state or action expressed by the verb.

– Factitive: the case of an object or being that results from the action or the state expressed by the verb or the action understood as part of the meaning of the verb.

– Locative: the case that identifies the location or spatial orientation of the state or the action expressed by the verb.

– Objective: the case that represents everything that is representable by a noun whose role in the action or state expressed by the verb depends directly on the semantic interpretation of the verb itself. This primarily consists of things affected by the action of the verb. This term should not be confused with the object complement. The sentence is analyzed in two constituents according to this rule: sentence → modality + proposition. Thus, the tree diagram of a sentence has the diagram form presented in Figure 2.5.

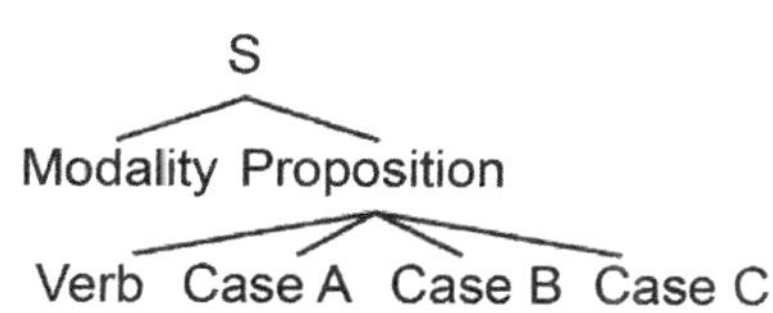

Figure 2.5. *Typical structure of a sentence according to Fillmore's model*

According to this approach, the modality includes information about the verbal predicate such as the time, mode, and type of sentence: affirmative, interrogative, declarative, negative, and the aspect, the representation of the locutor of the action expressed by the verb (see the example in Figure 2.5).

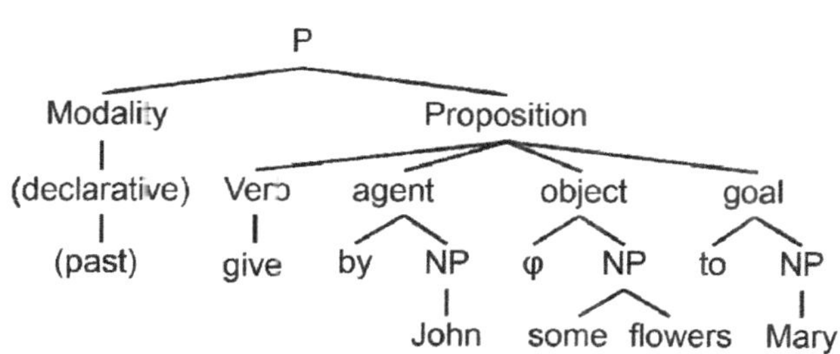

Figure 2.6. *Deep structure of the sentence John gave flowers to Mary*

The proposition represents the relation of the verb with one or more nominal groups in an atemporal manner (see examples in Figure 2.7).

John repaired the television with a screwdriver.

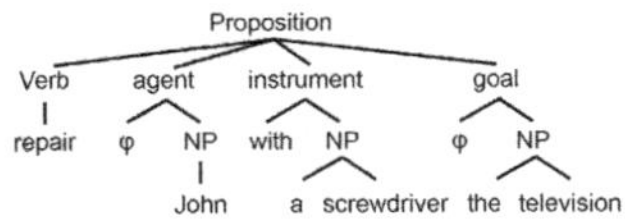

A screwdriver repaired the television.

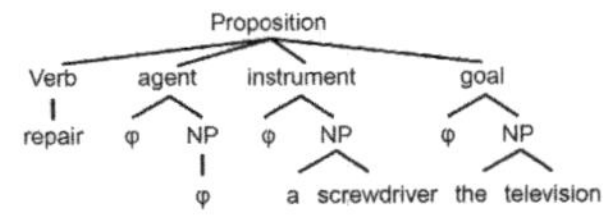

The television is repaired.

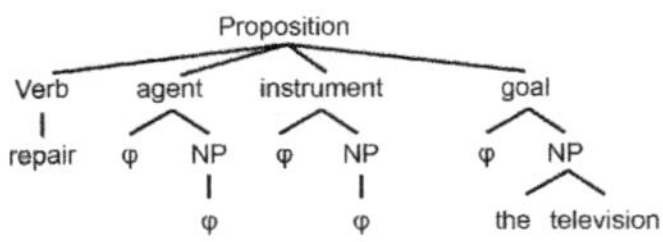

Figure 2.7. *Representations of the sentence: John repaired the television with a screwdriver*

During the 1970s and 1980s, Fillmore's work on case grammar gave rise to a new theoretical framework called frame semantics. This theory combined linguistic semantics and encyclopedic knowledge. It influenced the work of Ronald Langacker on cognitive grammar [LAN 87]. Other researchers, notably including Walter Cook, John Anderson and Lachlan Mackenzie, also worked on the development and application of case grammar [COO 71, COO 89, AND 77, MAC 81].

2.1.4. *Rastier's interpretive semantics*

Founded by François Rastier, a distinguished student of Greimas and Pottier, this new version of interpretive semantics was perceived by many experts as a new synthesis of structural semantics. It was developed in the wake of the works of European linguists Saussure, Hjelmslev, Greimas and Pottier [HEB 12].

Before presenting Rastier's semantic approach, it is important to first present the levels of linguistic description that he identifies in order to clarify his terminology and situate his work within the general context:

– The level of the morpheme: two types of morphemes are identified. Grammatical morphemes constitute a closed class in a given synchronous state.

For example, the plural marker *-s* as in *houses*, the inflectional morpheme ending *-ed* as in *walked*. Lexical morphemes (lexemes) belong to one or more loosely closed classes.

– The level of the lexia: the lexia is a group of integrated morphemes that constitute a unit of meaning. This includes simple lexia that are composed of a single morpheme and complex lexia that are composed of several morphemes. When they are written in several words, complex lexia can be classified into two groups: those that do not allow for the insertion of morphemes (maximal integration/total fixation) like *in single file*, and those that do allow for insertions (partial fixation) like *to step up to the plate, to step up quickly to the plate* [RAS 05a]. Note that, according to Rastier, the lexia is the most strongly integrated syntactic unit. Composed of one or more sememes, the signified of a lexia is a semia. It can be composed of one or more sememes.

– The level of the syntactic unit: considered to be the true place of predication, according to Rastier, the syntactic unit is the most important for morphosyntactic constraints.

– The level of the period: Rastier prefers to use the word period rather than the word sentence, which he considers to be a normative ideal created from the connection between grammar and logic. Notably elaborated by [CHA 86, BER 10], the period is the unit above the syntactic unit and its limits are rhetorical rather than logical. In speech, it is a respiratory unit. Generally, in speech and writing, it is a segment that can be defined using privileged local semantic relations between syntactic units (e.g. the phenomena of anaphora and coreference). The period defines the first level of a hermeneutic whole [RAS 94].

– The level of the text: this is a higher level of complexity on which other levels depend. This level is governed by discursive and stylistic norms[6]. According to Rastier, the semantic structure of texts can be described using a three-tiered model: microsemantics, the level of the morpheme and lexia; mesosemantics, the level of the syntactic unit and the period; and macrosemantics, the level of units above the period all the way up to text.

Microsemantics are the semantics of the level below the text. Its *upper* limit is the semia (the signified of a lexia). Three sections can be distinguished in microsemantics: semes, lexicalized units and contextual relations [RAS 05a].

A sememe is a structured set of features that are deemed pertinent, called semes. There are two types of semes: generic semes and specific semes.

6 According to Rastier, norms are less prescriptive than rules.

Generic semes are inherited from higher classes in the hierarchy (hyperonyms). They make it possible to mark relations of equivalence between the sememes. Rastier proposes distinguishing between three types of generic semes:

– Microgeneric: semes that indicate that a sememe belongs to a taxeme. According to Rastier, the taxeme is a minimal class of sememe. For example, *bus* and *metro* belong to the taxeme of /urban transportation/ and *coach* and *train* belong to the taxeme of /inter-urban transportation/. Taxemes are a minimal class of semia. For example, /funerary monument/ for "mausoleum" and "memorial".

– Mesogeneric: semes that indicate that a sememe belongs to a domain. The domain is the class of the level above the taxeme. It is related to a social practice. For example, *canapé* can refer to the domain //house// or the domain //food//.

– Macrogeneric: semes that indicate that a sememe belongs to a dimension. The dimension is a class of sememes or semia, of a higher level of generality, independent of domains. For example: //animate//, //inanimate//, or //human//, //animal//.

To illustrate this concept, Rastier studied the following four semias in the domain of transportation: *train*, *metro*, *bus* and *coach*. Two analyses are possible for these semias (see Table 2.1).

First analysis		Second analysis	
//transportation//		//transportation//	
//rails// "metro" /intra-urban/ "train" /extra-urban/	//road// "bus" /intra-urban/ "coach" /extra-urban/	//intra-urban// "metro" /rails/ "bus" /road/	//inter-urban// "train" /rails/ "coach" /road/

Table 2.1. *Analyses of semias: train, metro, bus and coach*

Although they are semantically valid, each of these analyses is preferred in a given context: in a technical context, the first analysis tends to be preferred, whereas in daily life, the second system is used. Finally, it is important to note that the generic semes of a sememe form its classeme.

In turn, specific semes serve to oppose a sememe to one or more other sememes in the same taxeme. For example, zebra is opposed to donkey by the (specific) seme /striped/. Similarly, mausoleum is opposed to memorial by the seme /presence of a body/. The specific forms of a sememe form a semanteme.

Both of these types of semes can have two different statuses that characterize their modes of actualization: inherent semes and afferent semes.

Inherent semes make it possible to define the type. They are inherited by default if the context does not forbid it. For example, /yellow/ is an inherent seme to banana because the typical color of a banana is yellow. Although it is inherited by default, this value can be changed by a different contextual instantiation such as: *Patrick painted a purple banana*. Thus, no inherent seme manifests in all contexts.

There are two kinds of afferent semes: socially normed semes and contextual semes. Socially normed semes indicate paradigmatic relations that can be applied from one taxeme to another. As they do not have a defining role (unlike inherent semes), these semes are normally latent. It is therefore only possible to actualize them through a contextual instruction. If, when talking about a shark, we evoke its extraordinary capacity to catch and kill its prey, the emphasis is therefore on the seme /ferocious/. The same applies when qualifying a businessman as a *shark*. Because contextual semes are not proper to the lexical item, they are transmitted through determinations or predications. The specificity of these semes is that they only involve the relations between occurrences, not considering the type. For example, in *the domesticated zebra*, the seme /domesticated/ must be represented in the occurrence of zebra.

Three interpretive operations about semes are proposed by Rastier to account for the transformations of significations encoded in languages [RAS 05a]: activation, inhibition and propagation. Governed by the laws of dissimilation and assimilation, these operations make it possible to increase semantic contrasts.

Inhibition is an operation that consists of blocking the actualization of inherent semes. Inhibited semes are therefore virtualized. For example, the inherent feature /animal/ of the word *horse* is actualized in the contexts such as *John rides his horse every morning*. In contrast, the same feature is inhibited in nominal locutions like: *Trojan horse, get off your high horse, dark horse* or *straight from the horse's mouth*.

Activation is an operation that allows the actualization of semes. Since inherent semes are actualized by default, activation only applies to afferent semes. For example, the seme /upright/ is not an inherent seme for the word *shepherd*. It is a virtual seme that can be inferred from the inherent seme /human/. In the context of O Eiffel[7] Tower shepherdess, by assimilation, the seme /upright/ is actualized by the presence of the inherent seme of tower: /verticality/.

7 Extracted from a poem by the French poet Guillaume Appolinaire.

The operation of propagation concerns contextual afferent semes given that the propagation of generic features occurs more naturally than specific features, which require particular contexts like metaphors or similes. For example, the word *doctor* entails neither the seme /meliorative/ nor the seme /pejorative/. However, in a context such as: *John is not a quack, he's a doctor*. Opposed with the word *quack*, which has /pejorative/ as an inherent feature, the word *doctor* acquires the feature /meliorative/. It is also useful to note that proper names of people lend themselves well to the propagation of semes, as their signifieds entail very few inherent semes. For example, characters in literary works represented by their proper names. Thus, Sherlock Holmes, the well-known character created by Sir Arthur Conan Doyle, only has two inherent features: /human/ and /masculine/. In the context of the novels, he receives the feature /strong observation/ and /high logical reasoning/.

As emphasized by Rastier, the interpretive operations just described are only applicable if certain conditions are met. Thus, to start an interpretive process, it is necessary to distinguish between the problem that it makes it possible to solve, the interpretant that selects the inference to be conducted, and the reception conditions that allow or facilitate the process. For example, some processes are facilitated within the same syntactic unit.

It should be noted that Rastier, based on his study of the possibilities of applications of the operations, proposes these three general principles. First, all semes can be virtualized by context. Second, only context makes it possible to determine whether a seme can be actualized or not. Third, there is no seme (inherent or not) that is actualized in all possible contexts.

Mesosemantics concerns the level between the lexia and the text. Thus, it pertains to the space that extends from the syntactic unit to the complex sentence [RAS 05b]. Although it is central in most linguistic studies, according to Rastier, syntax plays a secondary role in the domain of interpretation because he considered that ultimately, it was the hermeneutic order that took precedence over the syntagmatic order that includes syntax: semantic relations connect all sentences to their situational contexts.

Initially proposed by [GRE 66] to account for the homogeneity of discourse, the concept of isotopy is used by Rastier to illustrate relations that are deemed relative between syntax and semantics. It consists of the repetition of a seme, called an isotopizing seme, from one signified to another occupying a different position. For example, consider the text fragment [2.8]. In this fragment, the isotopy /urban transportation/ is formed by the repetition of the seme of the same name with the words *taxi*, *driver* and *boulevard* that possess this seme:

The taxi is driving 100 km/h down boulevard St. Germain.
The driver must be crazy... [2.8]

Following the status of the semes that they imply, Rastier distinguishes two types of isotopies: generic isotopies and specific isotopies.

Generic isotopies are divided into three sub-types:

– A microgeneric isotopy is marked by the recurrence of microgeneric semes. For example, the feature /telecommunications/ in *telephone* and *cell phone* and the feature /equine/ in *a sorrel horse*.

– A mesogeneric isotopy is defined as the recurrence of a mesogeneric seme. For example, the feature /writing tool/ in *a pen and a pencil*.

– A macrogeneric isotopy concerns the recurrence of a macrogeneric seme. For example, the feature /animated/ in *beauty and the beast*.

Specific isotopies can index sememes belonging to the same domain or to the same dimension, like tiger and zebra.

As noted by Rastier, isotopies also play a fundamental role in anaphors. As these only have a small number of inherent features, most of their afferent features are naturally propagated by context. This propagation is selective and only concerns part of the features of an anaphorized unit. It consists of phenomena that also concern the mesosemantic level as much as the macrosemantic level, because many cases of anaphors can surpass the framework of the sentence or the period.

According to Rastier's approach, the issue of relations between syntax and semantics can be reduced to relations between isosemy, prescribed by the functional system of language, and facultative isotopies, prescribed by other systems of norms [RAS 05b]. By limiting himself to facultative generic isotopies that index sememes and semias belonging to a single semantic domain, Rastier obtained these five configurations:

– Neither facultative isotopy nor isosemy: sequences that are neither sentences nor utterances. For example: *I slowly school the days mornings factory*.

– Isosemies, but not facultative isotopy: statements that are syntactically correct but for which no valid semantic interpretation exists, such as: *the Manichean chair travels through illegal yellow concepts*.

– Two interlaced domain isotopies, for example: *O Eiffel Tower shepherdess today your bridges are a bleating flock*[8] (French poet Apollinaire). The utterance

8 Translation by David Lehman http //www.vqronline.org/translations/apollinaires-zone

creates a complex referential impression, and it is undecidable because we cannot say whether it is true or false.

– A facultative isotopy, but with a rupture of isosemies, for example: *the gun furiously searches for its bullets while aiming at the inattentive game*. The words *gun, bullets* and *game* make up a seme that indexes them in the domain /hunting/. However, the world to which this utterance refers is both counterfactual and logically false.

– A facultative isotopy and isosemies: this includes decidable utterances where several sememes are indexed in a single domain. *John, a first-rate runner, won his first Olympic medal.*

Thus, for interpretive semantics, the interpretability of an utterance depends more on its facultative isotopies than its obligatory isotopies. Morphosyntax has a secondary place in the domain of interpretation.

Macrosemantics concerns the level of units above the period up to the text. Rastier considered that the production and interpretation of texts was the result of the non-sequential interaction thematic, dialectic, dialogic and tactic (see Figure 2.8).

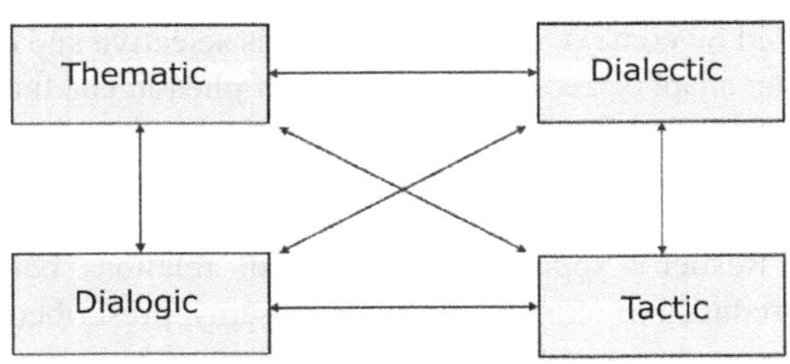

Figure 2.8. *Interactions of the independent components of macrosemantics*

The following paragraphs include a brief presentation of each of these components.

The thematic concerns the content of the text by specifying the topics that it addresses. According to Rastier, two types of topics can be distinguished: generic topics and specific topics.

Generic topics, or generic isotopies, are defined by a seme or a structure of generic semes. This recurrence defines an isotopy or isotopic bundle [RAS 05a]. Generic topics can take several forms such as taxemic topics, which pertain to members of the same taxeme. For example, a text that contains occurrences of *cargo*, *boat* and *ship* addresses the topic of //means of maritime transportation//. Similarly, there are domain topics (pertaining to members of the same domain), dimensional topics (pertaining to members of the same dimension) and semantic field topics.

Specific topics, also known as semic molecules, are recurring groups of specific semes. They can be presented in the form of a microsemantic graph without being actualized in lexical form. Rastier provides the example in Emile Zola's novel *L'Assommoir*[9] of a semic molecule that groups the semes /yellow/, /hot/, / viscous / and / harmful /, which are lexically realized by *alcohol*, *sauce*, *mucus*, *oil* and *urine*. Note that, in technical texts, semic molecules are generally lexicalized by terms.

Accounting for represented intervals of time, the dialectic is defined in two levels: the event level and the agonistic level. It allows for a dialectic typology and provides more detail about the concept of the narrative text[10].

The event level has three basic units: actors, roles and functions. An actor is the aggregation of the anaphoric actants of periods (mesosemantic unit). In the period, the actants can be named or have various descriptions. Each denomination or description lexicalizes one or more of the actor's semes. Roles are afferent semes that include the actant. They consist of semantic cases associated with the actants (accusative, ergative, dative, etc.). Functions are typical interactions between actors. They are defined by a semic molecule and generic semes. For example, the *gift* is an irenic function whereas the *challenge* is a polemic function. Each function has its own actor valence. For example, a [CHALLENGE] function can be written as: [A]<—(ERG)<—[CHALLENGE]—>(DAT)—>[B], where [A] and [B] are the actors.

Note that the semic relations between actants and functions explain the concordance or rection relations that are established between them.

Hierarchically superior to the event level, the agonistic level only appears in mythic texts and essentially only applies to technological texts. Two basic units can be identified in this level: agonists and sequences. An agonist is a type of a class of actors often indexed on different isotopies. For example, in *Toine*[11] by Maupassant, an old woman (an actor of human isotopy), the rooster (an actor of animal isotopy), and death (an actor of metaphysical isotopy) are metaphorically compared (see [RAS 89]). *Sequences* are defined by the homologation of functional

9 https://en.wikipedia.org/wiki/L%27Assommoir

10 According to Rastier, the narrative text is a relation of events involving people or characters.

11 http://www.classicshorts.com/stories/Toine.html

isomorphic syntactic units. For example, in *Dom Juan*[12] by Molière, there are 118 functions that can be grouped into only eleven sequences (see [RAS 73]). The sequences are ordered by logical narrative relations that are not necessarily chronological.

The dialogic accounts for the modalization of semantic units at all levels of complexity of the text. It concerns aspects of texts, especially literary texts, such as the universe, worlds and narration.

This component pertains to the linear order to semantic units at all levels. Although this order is directly related to the linearity of the signifiers, it does not combine at all levels.

2.1.5. *Meaning–text theory*

Meaning–text theory is a functional model of language that was initially developed in the Soviet Union by [ZOL 67] and then developed at the Université de Montréal, notably by Igor MEL'CUK (see [MEL 97, MEL 98, POL 98], and for a presentation of its NLP applications, refer to [BOL 04]). This theory is based on three premises:

– Language is a finite set of rules that establish a many-to-many correspondence between an infinite, uncountable set of meanings and an infinite, uncountable set of linguistic forms that are texts. This premise can be diagrammed like this:

{$RSem_i$} **language**;<==>; {$RPhon_j$} | 0 < i, j ∞

where *RSem* is a semantic representation and *RPhon* is a phonetic representation of the text.

– The input of a meaning–text model is a semantic representation and its output is the text. This correspondence must be described by a logic device that constitutes a functional model of language.

– The logic device of the second premise is based on two intermediary levels: words and sentences. These two intermediary levels correspond to morphological and syntactic patterns, respectively.

12 https://en.wikipedia.org/wiki/Dom_Juan

As shown by the architecture of the meaning–text model presented in Figure 2.9, the input of this model is a semantic representation. Getting to an output that is a corresponding sentence in phonetic transcription requires going through intermediary syntactic and morphological representations.

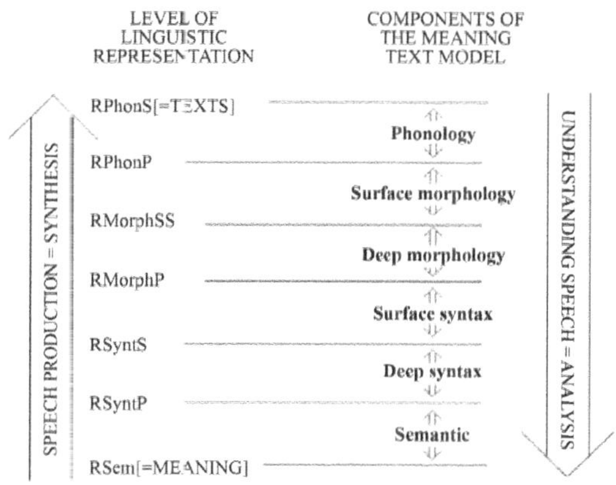

Figure 2.9. *Architecture of the meaning–text model [MEL 97]*

In Figure 2.9, the meaning–text model is a functional approach that describes language according to its mode of use: comprehension or production. Meaning–text correspondences are described by functions, in the mathematical sense, that establish a link between a meaning and all of the texts that express it. To describe an utterance, six levels of intermediary representation are used in addition to the levels of meaning and text. The articulation of a surface level and a deep level for the morphology and syntax serves to distinguish what is directed toward meaning from what is directed toward the text. The representations are made of formal objects called structures. The focus of this work is limited to the semantic level, which interests us the most in this context.

The semantic representation (RSem) includes three structures: the semantic structure, the semantic-communicative structure and the rhetorical structure.

The semantic structure and core of RSem corresponds to propositional meaning. It is a network whose nodes are labeled by semantic categories that are senses of the lexias of the language. The arcs of the network are labeled by numbers that specify the arguments of the predicate. To write the semantic structures, a formal language in the form of a network is used where the arcs indicate the predicate argument relations. For example, the predicate p(x, y) is represented with the network in Figure 2.10.

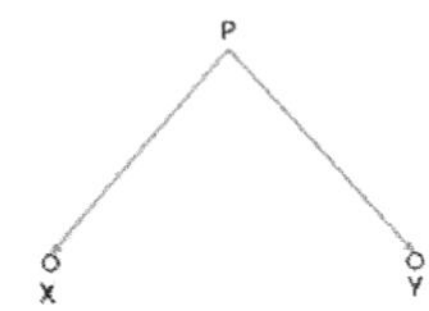

Figure 2.10. *Semantic network of the predicate p(x, y) [MEL 97]*

The semantic component is a set of rules that ensures correspondence with the deep level of syntactic representation (RSyntP).

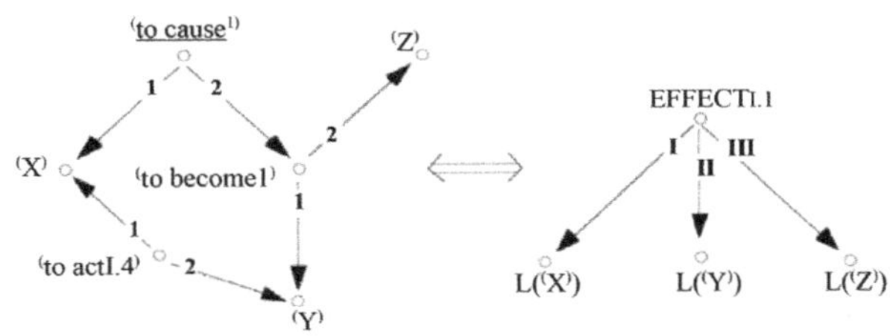

Figure 2.11. *Lexical semantic rule R1 [MEL 97]*

In Figure 2.11, the notation L(α) signifies that the lexia *L* expresses the meaning α. The rule R1 stipulates that the meaning (X, acting on Y, causes Y to become Z) can be expressed by the lexeme EFFECTI.1: *the effect Z of X on Y*. It should be noted that the lexical rules are the core of entries in a new type of dictionary: the explanatory combinatorial dictionary [MEL 84].

The semantic-communicative structure expresses the oppositions between the following pairs: topic and neutral, given and new, and emphatic and neutral. Finally, the rhetorical structure concerns the locutor's intentions such as irony or pathos.

2.2. Formal semantics

The idea of using mathematics as a universal framework for human thinking goes back at least to Descartes. Mathematics has the advantage of being systematic and clear due to the absence of ambiguities, the main problem of natural languages. In the same vein, the British logician and philosopher Bertrand Russell published his book *Principia Mathematica* in three volumes in 1910. Written in collaboration with Alfred North Whitehead, this work developed a symbolic language designed to avoid the ambiguities of natural languages. In turn, the German–American philosopher Rudolf Carnap also contributed by developing a symbolic logic that used mathematical notation to indicate the signification of signs unambiguously. This logic is, in a way, a formal language that serves as a metalanguage designated as the semiotics of this language. In this semiotics, each sign has a truth condition. When this condition is satisfied, the meaning of a sign corresponds to what it designates.

As noted in [BAC 89], logical semantics is based on two premises. The first premise, identified by Chomsky, is that linguistics can be described as a formal system. The second premise, formulated by [MON 70], is that natural languages can be described as interpreted formal systems.

Logical semantics was not unanimously accepted by linguists. Among others, it was criticized by François Rastier, who believed that it did not treat the linguistic signified but rather related it back to a logical form [RAS 01]. However, the existence of computer languages that adopt a logical framework like Prolog makes formal approaches an attractive choice for the computational processing of meaning.

Several types of logic were used to represent meaning. This presentation will examine the two main types: propositional logic and predicate logic.

2.2.1. *Propositional logic*

2.2.1.1. *The concept of proposition*

Propositional logic concerns declarative sentences that have a unique truth value (true or false). Sentences are the smallest logical unit that cannot be further broken down (see [DOY 85, LAB 04, AMS 06, GRA 04, ALL 01, KEE 05] for more detailed introductions to propositional logic). Consider the following sentences [2.9]:

a) Sydney is the capital of Australia.

b) The lowest point on earth is located near the Dead Sea. [2.9]

c) The Euphrates is the longest river in the world.

d) John will defend his thesis on June 12th.

e) No, thank you.

f) One kilo of tomatoes, please.

g) That's not exactly true.

In the series of examples in [2.9], sentences a–d are propositions that have specific truth values even if, at a given moment, certain values are not known as in the case of *d* (the date of his defense can change at any time). Sentences *e* and *f* are not propositions because they are not declarative and do not have a truth value. Sentence *g* can have two truth values at the same time, which is why it is not a proposition.

2.2.1.2. *Logical connectives*

Propositions only correspond to one particular type of sentence: simple sentences. For more complex forms of sentences, specific operators called logical connectives must be used. There are five of these connectives: negation, conjunction, disjunction, implication and biconditional. These connectives act as operators on the truth values of propositions.

In propositional logic, the negation of a proposition consists of inverting its truth value. Thus, the proposition ¬*P* (sometimes written as ~*P* and read as not *P*) is true if *P* is false. Inversely, it is false if *P* is true. This gives us the truth table in Table 2. 2.

P	¬P
T	F
F	T

Table 2.2. *Truth table for the negation operator*

Note that the variable *P* can correspond to a simple proposition or a more complex proposition like ¬(Q ∧ R). Unlike natural languages where negation can concern several constituents, in propositional logic, negation can only pertain to the whole proposition. For example, in the sentence, *John did not study at the faculty of medicine in Paris last year*, the negation concerns three parts of the sentence that convey different information: *study*, *at the faculty of medicine in Paris* and *last year*. Sometimes, propositional negation corresponds to a total negation (that concerns the entire sentence). *John is not agreeable* is the equivalent of: ¬ (John is agreeable). This kind of negation can be realized with a morphological prefix: *in-*, *im-*, *il-* and *ir-*. For example, *The President of the Republic is ineffective* is equivalent to ¬ (The President of the Republic is effective).

Conjunction (∧) relates two propositions, *P* and *Q*, whose truth values are given in Table 2.3. It is sometimes written as (P & Q) and is read as: *P* and *Q*.

P	Q	P∧Q
T	T	T
F	F	F
T	F	F
F	T	F

Table 2.3. *Truth values for the conjunction operator*

Assuming that the propositions *John saw the painting*, *John liked the painting* correspond to propositions *P* and *Q* respectively, then the following sentences can be formulated logically in this way:

a) John saw the painting and he liked it. P ∧ Q

b) John saw the painting but he didn't like it. P ∧ ¬Q

c) John saw the painting. However, he didn't like it. P ∧ ¬Q

d) John liked the painting although he had not seen it. Q ∧ ¬P

Note that the logical equivalence between (P ∧ Q) and (Q ∧ P) is always valid. However, there are cases where this equivalence is not valid in natural language. For instance, consider the example: John opened the door and he left his house. The change in the order of propositions here implies a semantic change as the idea of coordination is paired with the idea of a chronological succession of events expressed by the propositions. With verbs that express reciprocity, like *to date*, *to marry*, *to affiance* and *to love*, the coordination in natural languages does not translate by using a logical *and*.

Disjunction (V / VV) of two propositions P V Q, which is read as P or Q, is true if at least one of the propositions is true. This operator is called inclusive disjunction. Its truth table is provided in Table 2.4.

P	Q	P V Q
T	T	T
F	F	F
T	F	T
F	T	T

Table 2.4. *Truth table for the or operator*

Inclusive disjunction is observed in sentences like:

John or Mary go to school.	(John goes to school) V (Mary goes to school)
John works in the library or at home.	(John works at the library) V (John works at home)

There is a variant of disjunction called exclusive disjunction (noted as **VV** or XOR) that is only true if one of the two propositions is true: it is false if the two propositions are true or false. P **VV** Q is read as P or Q but not both. See Table 2.5 for the truth table of this variant.

P	Q	P VV Q
T	T	F
F	F	F
T	F	T
F	T	T

Table 2.5. *Truth table for the exclusive or operator*

In daily life, the *exclusive or* designates an action that only takes a single agent as in *John or Michael is driving Mary's car*. This also translates into expressions such as *Either John or Mary is leaving on a mission to Paris*.

Implication is represented by the operator →. If *P* and *Q* are propositions, then *P* → *Q*, is read as: *if P then Q* or *P implies Q*, is a proposition whose truth table is provided in Table 2.6.

	P	Q	P → Q
1	T	T	T
2	F	F	T
3	F	T	T
4	T	F	F

Table 2.6. *Truth table for the implication operator*

In Table 2.6, P → Q is false only if P is true and Q is false. To clarify this table, consider the four sentences below, which correspond to the four lines in the truth table, respectively:

1) If it is nice out, I will go play sports in the park.

2) If London is the capital of the Maldives, then Paris is the capital of Sri Lanka.

3) If you are Marie Antoinette's friend, then the Earth is round.

4) If Damascus is the capital of Syria, then 8 + 5 = 20.

Propositions 2 and 3 indicate that anything can be deduced from something that is false. Proposition 4 is false because something false cannot be deduced from something that is true. Note that more complex propositions can be part of an implication as in this sentence: if it is nice out (P) and if I am not tired (¬Q), then I will go play sports (R).

$$(P \land \neg Q) \rightarrow R$$

The biconditional, represented by the operator ↔, expresses a relation of equivalence between two propositions. The proposition P ↔ Q (which is read as P if and only if Q) is true if the two propositions P and Q have the same truth. This results in the truth table provided in Table 2.7.

P	**Q**	**P↔Q**
T	T	T
F	F	T
T	F	F
F	T	F

Table 2.7. *Truth table for the biconditional*

From a logical perspective, P↔Q is equivalent to P → Q ∧ Q → P. Expressed in English, this relation is translated by expressions such as: *only if, without which, unless*. Consider the following sentences and their implications:

I will come to the game (p) unless I am on a trip. (Q).	P↔Q
If I am on a trip, then I will not come to the game.	Q → P
If I come to the game, then I am not on a trip.	P → Q

2.2.1.3. *Well-formed formulas*

A well-formed formula (wff) is a logical expression that has a meaning. A simple variable that represents a proposition *P* is the simplest form of a wff. Connectives can be used to obtain more complex wffs. For example: (¬P), (p ∧ ¬Q), (¬¬P), etc. More formally, a wff can be defined according to these criteria:

i) All atomic propositions are wffs.

ii) If *P* is a wff, then ¬*P* is a wff.

iii) If *P* and *Q* are wffs, then (P ∧ Q), (P ∨ Q), (P → Q), (P ↔ Q) are wffs.

iv) A sequence is a wff if and only if it can be obtained by a finite number of applications of (i)–(iii).

By replacing the variables of a wff α with equivalent propositions, the truth value of this formula can be calculated from the truth values of the propositions that compose it. Consider this formula and its truth table presented in Table 2.8:

$\alpha = (P \lor Q) \land (P \rightarrow Q) \land (Q \rightarrow P)$

P	Q	(P ∨ Q)	(P → Q)	(P → Q) ∧ (Q → P)	(Q → P)	α
T	T	T	T	T	T	T
T	F	T	F	F	T	F
F	T	T	T	T	F	F
F	F	F	T	F	T	F

Table 2.8. *Truth table for formula α*

Note that the number of lines in the formula increases with the number of variables in the formula. It is equal to 2^n, where n is the number of variables in the formula. This means that the number of lines in the truth table of a formula with two variables is equal to four, and for a formula with three variables, there are eight lines, and so on.

Some formulas have the truth value T, regardless of the value of the propositions that compose them. In other words, they are always true. This type of formula is called a tautology. The formulas P ∨ ¬P, (P ∧ Q) ∨ ¬P ∨ ¬Q are examples of tautologies. Another particular case of a wff is absurdity or contradiction. This is a wff that, unlike a tautology, has the truth value F, regardless of the truth value of the propositions that compose it. For example, P ∧ ¬P, (P ∧ Q) ∧ ¬Q. Because absurdity is the opposite of a tautology, α is an absurdity if ¬α is a tautology, and vice versa.

Two wffs α and β with propositional variables p_1, p_2, ..., p_n, etc. are logically equivalent if the formula $\alpha \leftrightarrow \beta$ is a tautology. Therefore, this is written as $\alpha \equiv \beta$. This means that the truth tables of α and β are identical. For example:

P ∧ P ≡ P, consider the truth table for the two formulas presented in Table 2.9.

P	P ∧ P
T	T
F	F

Table 2.9. *Proof of P ∧ P ≡ P*

Because the columns corresponding to the formulas are identical, this proves that the formulas are equivalent. To prove the equivalence: P ∧ Q ≡ Q ∧ P, consider the truth table for the two formulas presented in Table 2.10.

P	**Q**	**P∧Q**	**Q∧P**
T	T	**T**	**T**
T	F	**F**	**F**
F	T	**F**	**F**
F	F	**F**	**F**

Table 2.10. *Proof of P ∧ Q ≡ Q ∧P*

To clarify the concept of proof of equivalence, consider one final example, (p → (Q **V R**)) ≡ ((P → Q) **V** (P → R)) whose equivalence is proven in Table 2.11 where the two wffs have exactly the same truth values.

P	**Q**	**R**	**Q V R**	**p → (Q V R)**	**P → Q**	**P → R**	**(P → Q) V (P → R)**
T	T	T	T	**T**	T	T	**T**
T	T	F	T	**T**	T	F	**T**
T	F	T	T	**T**	F	T	**T**
T	F	F	F	**F**	F	F	**F**
F	T	T	T	**T**	T	T	**T**
F	T	F	T	**T**	T	T	**T**
F	F	T	T	**T**	T	T	**T**
F	F	F	F	**T**	T	T	**T**

Table 2.11. *Proof of (p → (Q ∨ R)) ≡ ((P → Q) ∨ (P → R))*

It should be added that there are particular equivalencies that serve to deduce other equivalencies by simplifying the formulas. These are called identities. If a formula β is part of a formula α and if β is equivalent to β', then it can be replaced with β'and the formula obtained is a wff equivalent to α. Figure 2.12 provides a few examples of logical identities commonly used in deductions.

I_1 Laws of commutativity

$P \vee Q \equiv Q \vee P$

$P \wedge Q \equiv Q \wedge P$

I_2 Laws of associativity

$P \vee (Q \vee R) \equiv (P \vee Q) \vee R$

$P \wedge (Q \wedge R) \equiv (P \wedge Q) \wedge R$

I_3 Laws of distributivity

$P \vee (Q \wedge R) \equiv (P \vee Q) \wedge (P \vee R)$

$P \wedge (Q \vee R) \equiv (P \wedge Q) \vee (P \wedge R)$

I_4 Laws of absorption

$P \vee (P \wedge Q) \equiv P$

$P \wedge (P \vee Q) \equiv P$

I_5 Laws of DeMorgan

$\neg(P \vee Q) \equiv \neg P \wedge \neg Q$

$\neg(P \wedge Q) \equiv \neg P \vee \neg Q$

I_6 Laws of double negation

$P \equiv \neg(\neg P)$

I_7

$P \vee T$ (Tautology) $\equiv T$

$P \wedge T \equiv P$

I_8

$P \vee A$ (Absurdity) $\equiv P$

$P \wedge A \equiv A$

Figure 2.12. *A few logical identities*

To illustrate the use of these rules in the proof process, consider the following example: $(P \wedge Q) \vee (P \wedge \neg Q) \equiv P$. The proof consists of simplifying the left side until *P* is obtained.

$$(P \wedge Q) \vee (P \wedge \neg Q) \equiv P \wedge (Q \vee \neg Q). \quad I_3$$

$$\equiv P \wedge T \quad I_7$$

$$\equiv P$$

2.2.1.4. *Rules of inference*

Rules of inference are tautologies that have this construction: $\frac{\text{Premises}}{\therefore\text{Conclusion}}$

They are similar to an implication: Premises → Conclusion. The part that concerns the premises pertains to the proposition that is assumed to be true. It is sometimes called the hypothesis. The conclusion concerns the proposition derived from the premises. The valid process argument concerns arriving at the conclusion from the premises. Some deduction rules and their equivalents in the form of inferences are given in Figure 2.13.

Inference rule		Form of implication
	RI_1: **Addition**	
$\frac{P}{\therefore P \lor Q}$		$P \rightarrow (P \lor Q)$
	RI_2: **Conjunction**	
$\frac{P \quad Q}{\therefore P \land Q}$		$P \land Q \rightarrow P \land Q$
	RI_3: **Simplification**	
$\frac{P \land Q}{\therefore P}$		$(P \land Q) \rightarrow P$
	RI_4: **Modus ponens**	
$\frac{P \quad P \rightarrow Q}{\therefore Q}$		$(P \land (P \rightarrow Q)) \rightarrow Q$
	RI_5: **Modus tollens**	
$\frac{\neg Q \quad P \rightarrow Q}{\therefore \neg P}$		$(\neg Q \land (P \rightarrow Q)) \rightarrow \neg Q$
	RI_6: **Hypothetical syllogism**	
$\frac{P \rightarrow Q \quad Q \rightarrow R}{\therefore P \rightarrow R}$		$((P \rightarrow Q) \land (Q \rightarrow R)) \rightarrow (P \rightarrow R)$

Figure 2.13. *A few rules of inference*

Consider the following argument:

If John won an Olympic medal or if he won a gold medal at the world championships, then he is certain to receive the Legion of Honor[13]. If John is certain to receive the Legion of Honor, then he is happy. Or, he is not happy. Then, he did not win a gold medal at the world championships.

To verify the validity of such an argument, follow these steps.

1) Identify the propositions:

– P signifies: John won an Olympic medal.

– Q signifies: John won a gold medal at the world championships.

– R signifies: John is certain to receive the Legion of Honor.

– S signifies: John is happy.

2) Determine the left and right sides of the rule of inference:

The premises of the inference are:

i) $(P \vee Q) \rightarrow R$

ii) $R \rightarrow S$

iii) $\neg S$

The conclusion is: $\neg Q$

i) $(P \vee Q) \rightarrow R$	Premise (i)
ii) $R \rightarrow S$	Premise (ii)
iii) $(P \vee Q) \rightarrow S$	Hypothetical syllogism RI_6
iv) $\neg S$	Premise (iii)
v) $\neg(P \vee Q)$	Modus tollens RI_5, lines 3 and 4.
vi) $\neg P \wedge \neg Q$	Law of DeMorgan
vii) $\neg Q$	Simplification RI_3

3) It can be concluded that the argument is valid.

13 National Order of the Legion of Honour or *Legion of honor* is the highest French order of merit for military and civil merits.

In short, propositional logic is a declarative system that makes it possible to express partial knowledge, disjunctions, negatives, etc. However, contrary to natural languages, in propositional logic, meaning is independent of the context. Thus, each fact to be represented requires a separate proposition. This means that the representation process is not very efficient and sometimes insufficient. Propositional logic is incapable of adequately processing the properties of individual entities such as the Mona Lisa and Picasso or their relations to other entities. In addition, cases like [2.10] present difficulties for the logic of propositions because they require a mechanism that is not available in this type of logic. This concerns quantifiers that are available in the upper level logics like first-order logic:

– Everyone likes nature. [2.10]

– There is always someone to love a child.

2.2.2. *First-order logic*

First-order logic makes it possible to represent the semantics of natural languages in a more flexible and compact way than propositional logic. Contrary to the logic of propositions, which assumes that the world only contains facts, this logic assumes that the world contains terms, predicates and quantifiers.

2.2.2.1. *Terms*

To refer to individuals, first-order logic uses terms. These are specific objects that can be constant, like 10, Paul, and Lyon, or variables such as X, Y, Z. The functions of the form $f(t_1,...,t_n)$ are terms whose parameters t_i are also terms. For example, the function *prec(X)* is a function that takes an integer and gives the arithmetically preceding integer $(prec(X) = X-1)$. Another example of a function with several parameters is addition $+(X, Y)=X + Y$.

2.2.2.2. *Predicates*

Predicates are used to describe terms or to express the relations that exist between them. This is how predicate logic expresses propositions. Generally, unary predicates, with a single argument, can represent the properties or simple actions that can be true or false. For example: intelligent (John), left (John) and carnivore (tiger). N-ary predicates, with *n* arguments where *n* is greater than 1, represent predicates with several arguments. For example: the predicate mother (Mary, Theresa) can be used to designate Mary as the mother of Theresa or the inverse, with order being a question of convention. Similarly, the predicate: to leave (John, Cleveland, Chicago) can be interpreted as: *John left from Cleveland for Chicago* or *from Chicago for Cleveland*. Connectives can be used to express more complex facts, as in the sentences in [2.11] and their representations in first-order logic:

Theresa is the mother of John and Mary.

mother(Mary, Theresa) Λ mother(John, Theresa).

John likes oranges but he doesn't like apples.

like(John, Orange) ∧ ¬like(John, apple). [2.11]

Mary is studying pharmacy or medicine.

study(Mary, pharmacy) ∨ study(Mary, medicine).

2.2.2.3. *Quantifiers*

Quantifiers make it possible to express facts that apply to a set of objects, expressed in the form of terms, rather than on individual objects. To do the same thing in predicate logic, all of the cases must be listed, which is often impossible. For example, the sentences in group [2.12] require a representation with quantifiers:

– Everyone loves apple pie.

– Everyone is at least as poor as John. [2.12]

– Someone is far from home.

Predicate logic has two quantifiers to express this kind of sentence: the universal quantifier and the existential qualifier.

The universal quantifier has the form: x p(x). It is read as for all X p(x) and signifies that the sentence is always true for all values of the variable x. Consider examples [2.13]:

∀x likes(x, Venice)	Everyone likes Venice.	
∀x [horse(x)→ mammal(x) Λ mammal(x) →animal(x)]	Horses are mammals which are animals.	
∀x inherit(John, x) → book (x)	All that John inherited was a book.	[2.13]
∀x book(x) → inherit(John, x)	John inherited all of the books (in the universe).	

The universal quantifier is often used with implication. If not, it becomes very restricting. Consider this example: $\forall x$ apple(x) $\wedge$ red(x). This formula translates in natural language as: everything in the universe is a red apple.

The existential quantifier is expressed in the form $\exists x$ p(x) and is read as: there exists one x such as p(x) or there is at least one x such as p(x) and signifies that there is at least one value of x for which the predicate p(x) is true.

$\exists x$ bird(x) $\wedge$ in(forest, x)	There is at least one bird in the forest.
$\exists x$ mother(Mary, x) $\wedge$ mother(John, x)	John and Mary are siblings.
$\exists x$ person(x) $\wedge$ likes(x, salad)	There is (at least) one person who likes salad.
$\exists x(\forall y$ animal(y) $\rightarrow \neg$to like(x, y))	There is someone who does not like animals.
$\exists x \forall y$ like (y, x) $\wedge \neg\exists x \forall y$ like(x, y)	Everyone likes someone and no one likes everyone.
$\forall x(\exists y$(mother(y, x)) $\wedge \exists z$(y = father(z, x)))	Everyone has a father and a mother.

From a syntactic perspective, the negation connectives and the quantifiers have the highest priority. Then come the connectives of conjunction and disjunction. After that, implication, and finally, the biconditional has the lowest priority.

$\forall x$ cat(X) $\rightarrow$ nice(X).	All cats are nice.
$\exists x$ cat(X) $\wedge$ nice(X).	There is at least one cat that is nice.

Note that the same thing can be expressed with the existential quantifier and the universal quantifier. Consider these basic cases:

$\forall x \neg P \leftrightarrow \neg \exists x P$	
$\forall x \neg$ like(x, John) $\leftrightarrow \neg \exists x$ like(x, John)	Nobody likes John.
$\neg \forall x P \leftrightarrow \exists x \neg P$	
$\neg \forall x$ like(x, John) $\leftrightarrow \exists x \neg$ like (x, John)	There is at least one person who does not like John.
$\forall x P \leftrightarrow \neg \exists x \neg P$	
$\forall x$ like(x, John) $\leftrightarrow \neg \exists x \neg$ like(x, John)	Everyone likes John.
$\exists x P \leftrightarrow \neg \forall x \neg P$	
$\exists x$ like(x, John) $\leftrightarrow \neg \forall x \neg$ like(x, John)	There is at least one person who likes John.

There are some equivalencies that allow for making simplifications. For example:

$\forall x\ P(x) \wedge Q(x) \leftrightarrow \forall x P(x) \wedge \forall x Q(x)$

$\exists x\ P(x) \vee Q(x) \leftrightarrow \exists x P(x) \vee \exists x Q(x)$

To talk about a unique case or a particular entity that performs a given action or that has a given property, as in *only John makes/eats/sleeps*, the following formula can be used:

$\exists x\ P(x) \wedge \forall y\ P(y) \rightarrow (x = y)$. A particular version of the existential quantifier makes it possible to simplify the previous formula: $\exists! x\ P(x)$.

The use of a quantifier leads to the distinction between two types of variables: bound variables, which are found in the field of a quantifier, and free variables, which are independent from quantifiers. Sometimes, in the same formula, a variable can have both free and bound occurrences. Consider the following examples to clarify the difference between bound and free variables:

– $\exists x$ like(x, y) *x* is a bound variable, whereas *y* is a free variable.

– $\forall x \exists y(O(x, y)) \vee \exists z(P(z, x))$ the first occurrence of x is bound and the second occurrence is free. The variables *y* and *z* are bound.

The field of a variable makes it possible to describe some ambiguities. Consider the following sentence with its two possible paraphrases: each person (everyone) likes someone:

– Each person has someone that they like.

– There is a person that is liked by everyone.

The first interpretation implies a relation between two groups: several to several. On the contrary, the second interpretation of the sentence is a relation between a group and an individual: one to many. Translated logically, this provides the following formulas:

$\forall x \exists y$ like(x, y).

$\exists x \forall y$ like(y, x).

2.2.2.4. *Well-formed formulas (wffs)*

In first-order logic, the formulas are formally defined in this way:

i) The symbols of the predicate: if $P(a_1, ..., a_n)$ is a symbol of the predicate n-ary and $t_1, ..., t_n$ are the terms, the $P(t_1, ..., t_n)$ is a formula.

ii) Equality: if the equality symbol is considered to be part of the logic and t_1 and t_2 as being terms, then $t_1 = t_2$ is a formula.

iii) Negation: if φ is a formula, then $\neg \varphi$ is also a formula.

iv) Connectives: if φ and ψ are formulas, then: $(\varphi \wedge \psi)$, $(\varphi \vee \psi)$, $(\varphi \rightarrow \psi)$, $(\varphi \leftrightarrow \psi)$ are formulas.

v) Quantifiers: if φ is a formula, then $\forall x\ \varphi$ and $\exists x\ \varphi$ are formulas.

vi) Only the expressions obtained following a finite number of applications of (i)–(v) are formulas.

The formulas obtained with (i) and (ii) are called atomic formulas. For example, like(John, school) is an atomic formula whereas: go(John, school) $\rightarrow$ work(John) $\vee$ play(John) is a complex formula. The formulas can be analyzed in the form of trees made up of the atomic formulas that constitute them. Consider the following formula $p \rightarrow (q \vee r)$, analyzed in tree form as in Figure 2.14.

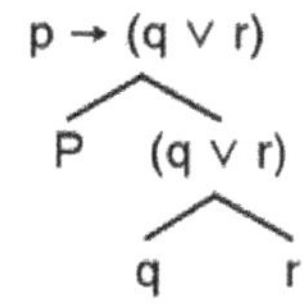

Figure 2.14. *Tree structure of a simple formula*

Consider another more elaborate example whose analysis is provided in Figure 2.15: $[(p \wedge q) \rightarrow (\neg p \vee \neg\neg q)] \vee (q \rightarrow \neg p)$.

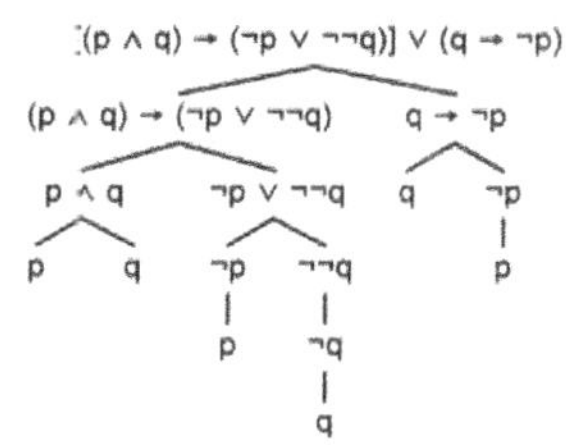

Figure 2.15. *Tree structure of a more complex formula*

Finally, it is useful to add that the formula, all of whose variables are bound, is called a closed formula or sentence whereas a formula with open variables is called a propositional function.

2.2.2.5. *Semantic interpretation*

Translating natural language sentences into a first-order logic equivalent is definitely a necessary step to reach a semantic representation of these sentences, but it is not sufficient. The logical representation must be connected to the context to express meaning. In principle, calculating the truth value of a sentence comes down to affirming or not whether it corresponds to the reality it describes. The truth value is positive or true (V = 1) in case of correspondence and negative or false (V=0) in case of non-correspondence. More formally, if the variable *S* is used to designate the situation described by the sentence, the sentence *P* is true in a situation *S* if $[p]^v=1$. Where $[p]^v$ is the truth value of the sentence *P*. Inversely, a sentence *P* is false in a situation *S* if $[p]^v=0$. The problem is that a sentence in natural language is composed of a set of elements such as the verb, subject nominal group, object complements, etc. To calculate the compositional meaning, the following steps are necessary:

i) Interpreting the first-order logic symbols: unlike in proposition logic, not everything in first-order logic is a proposition. Calculating the truth value of a sentence must begin by processing the basic symbols, each of which has an interpretation. The language of first-order logic is composed of the vocabulary *V* that is a set of constants, functions and predicates of this language.

ii) Defining the universe: sometimes called the domain, the universe is a representation of individuals and their relations in a given situation *S*. Consider, for example, the situation of the competition between David Douillet and Shinichi Shinohara for the Olympic medal in Judo in the +100 kg category in Sydney in 2000. To simplify it, the following individuals are in this situation: the two judokas, the judge and two spectators named Guillaume and Tadahiro. This gives a situation *S* with a set of individuals U={David, Shinichi, judge, Guillaume, Tadahiro}.

iii) An interpretation function: the role of this function is to establish a link between the logical representations and the extensions that correspond to these representations in the situation. In our situation *S*, the interpretation function *F(x)* can give the entity denoted by *x* in the situation *S*. For example:

F(d) = David

F(s) = Shinichi

F(g) = Guillaume

F(t) = Tadahiro

F(j) = judge

This function also provides the extensions of the predicates.

F(C) = **c**ombat in the Judo finals at the Sydney Olympics = {David, Shinichi}

F(M) = gold **m**edal winner in Judo at the Sydney Olympics = {David}

F(A) = **a**djudicated the Judo finals at the Sydney Olympics = {referee}

F(S) = **s**pectators at the Judo finals at the Sydney Olympics = {Guillaume, Tadahiro}

iv) The model: this is a combination of the domain and the interpretation function. From a formal perspective, this can be represented as: $M_n = (U_n, F_n)$ where *M* is the model, *U* is the set of individuals and *F* is the interpretation function. The specifiers make it possible to differentiate between the situations: $M_1 = (U_1, F_1)$; $M_2 = (U_2, F_2)$, etc.

v) Evaluating the truth values of the formulas: after having defined the general framework that makes it possible to anchor a formula in a given situation, it is necessary to define the rules (algorithms) that make it possible to decide if a formula is true or not in a given situation. Consider a simple expression like: S(ch).

In relation to the situation *S*, this expression signifies that Charles is a spectator at the +100 kg Judo finals at the Sydney Olympics. Intuitively, Charles is a spectator at this competition if the extension of Charles is part of the set of spectators defined by the predicate *S* (spectator) in the model M_1. The formula to evaluate if this expression is true or not is:

$[S(c)]^{M1} = 1$ iff[14] $[c]^{M1} \in [S]^{M1}$.

Because *Charles* does not belong to the set of spectators {Guillaume, Tadahiro}, then the expression *S(ch)* is false and its truth value is consequently equal to zero.

By taking the truth values of simple expressions as a base and considering a model *M* as a reference, the truth values of complex formulas can be calculated according to these rules:

– Identity: if φ expresses an identity: α = β, then φ is true in *M* iff β and α correspond to the same objects in *M*.

– Negation: if φ = ¬ψ, then φ is true in *M*, then ψ in *M* is false and vice versa.

– Equality: if φ = ψ, then φ is true if *φ* and *ψ* have exactly the same truth value in *M*.

– Conjunction: if φ = (α ∧ β), then φ is true in *M* iff α and β are true in *M*.

– Disjunction: if φ = (α ∨ β), then φ is true in *M* iff α or β are true in *M*.

– Implication: if φ = (α → β), then φ is false iff α is true and β is false.

– Biconditional: if φ = (α ↔ β), then φ is true in *M* iff α and β have the same truth value in M (both are true or both are false).

– Existential quantifier: if φ is in the form ∃x P(x), then its truth value will be true if there is u ∈U such as P(x) is true when *x* has the value *u*. In other words, φ is true if there is at least one value of x that is part of: $[x]^{M1} \in [P]^{M1}$. Returning to our model, M_1, the formula ∃x S(x) is true if *x* is equal to Guillaume or Tadahiro.

– Universal quantifier: if φ is in the form ∀x P(x), then it is true if *x* takes all of the possible values in U: ∀u ∈U, x=u.

2.2.3. *Lambda calculus*

Lambda calculus was invented in 1936 by Alonzo Church [CHU 40], at the same time that Alan Turing invented his machine. These two formalisms have the same

14 If and only if.

power of representation concerning computational calculations. Lambda calculus is a formalism just like predicate logic and first-order logic. In the domain of formal semantics, it is often considered to be an extension of first-order logic to include the operator lambda (λ) that makes it possible to connect variables. Although it initially lacked types, lambda calculus was quickly equipped with them. The types refer to objects of a syntactic nature. The non-typed variant is the simplest lambda calculus form where there is only one type. As noted in several books and tutorials entirely or partially dedicated to the subject, the use of lambda calculus became common in the domain of computational linguistics as well as in the domain of functional programming with languages like Lisp, Haskell and Python (for an introduction to lambda calculus, see [ROJ 98, MIC 89, BUR 04, BLA 05, LEC 06]). Due to its capacity to represent any computational calculation, lambda calculus lends itself well to the abstract modeling of algorithms [KLU 05].

2.2.3.1. *The syntax of lambda expressions*

The expression is the central unit of lambda calculus. A variable is an identifier that can be noted with any letter *a, b, c,* etc. An expression can be recursively defined like this:

<expression> := <variable> | <function>|<application>

<function> := λ<variable>.<expression>

<application> := <expression><expression>

The only symbols used by the expressions are the λ and a period. This gives expressions like:

λx.x	Identity function.
(λx.x)y	Application of a function to an expression.
λx.large(x)	An expression with a variable.
λxλy.eat(x)(y)	An expression with two variables.

The expression can be enclosed by parentheses to facilitate reading without affecting the meaning of this expression. If *E* is an expression, then the expression *(E)* is totally identical. That being said, there are different equivalent notations in the literature, where a period cannot be used, or in which square brackets are used to mark the extent of the lambda term. Consider the example of the predicate *eat* to illustrate these notations:

λxλy[eat(x)(y)]

λxλy(eat(x)(y))

λxy. eat(x)(y)

The applications of functions are evaluated by replacing the value of the argument *x* in the definition of the function: (λx.x)y = [y/x]x = y. In this transformation, [y/x] signifies that all of the occurrences of *x* are being replaced by *y* in the expression on the right. In this example, the identity function has just been applied to the variable y.

In lambda calculus, the variables are considered to be local in the definitions. In the function λx.x, the variable *x* is a bound variable because it is preceded by λx in the definition. On the other hand, a variable that is not preceded by λ is said to be a free variable. In the expression, λx.xy *x* is a bound variable while *y* is a free variable. A term that does not have a free variable is said to be closed or combinatory. More formally, a variable is free in these cases:

– <variable> is free in <variable>

– <variable> is free in λ<variable$_1$>.<expression> if <variable> ≠<variable1> and if <variable> is a free variable in <expression>.

– <variable> is free in E_1E_2 if it is free in E_1 or E_2.

Similarly, a variable is said to be bound in these cases:

– <variable> is bound in λ<variable$_1$>.<expression> if <variable> = <variable1> or if <variable> is a bound variable in <expression>.

– <variable> is bound in E_1E_2 if <variable> is bound in E_1 or E_2.

Note that sometimes, in the same expression, a variable can be both free and bound. For example, in the expression (λx.xy)(λy.yz), the first variable y is free in the first sub-expression (on the left) and it is bound in the second sub-expression (on the right).

2.2.3.2. *Types*

The use of the typed version of lambda calculus offers the advantage of avoiding paradoxes and keeping the function definition within the boundaries of the standard theory of sets. Without types, we could construct terms like: the set of all sets that are not self-included. This seemingly simple expression leads to a paradox known as the Russell Paradox. If this set is not a member of itself, then it is not the set of all sets and if it is a member of itself, then it was self-included.

According to the theory of types introduced into semantics by [MON 73], a type domain for a language *L* is defined by two elements *e* and *v* where *e* corresponds to an entity and *v* corresponds to a truth value (0 or 1). The derived types are defined in this way:

– *e* is a semantic type.

– *v* is a semantic type.

– For either of the semantic types δ, ε, $< \delta, \varepsilon >$ is a semantic type.

– Nothing else is a semantic type.

The domains are defined in the following way:

– D_e is the domain of entities or individuals.

– $D_{t=}$ {0, 1} is the domain of truth values.

– For either of the semantic types δ, ε, $D_{< \delta, \varepsilon >}$ is the domain of functions D_δ to D_ε.

According to this definition, the sentences (propositions) are type *v*, the nouns are type *e*. The other types are constructed based on these two types. Thus, intransitive verbs and adjectives, represented by predicates with a single argument, are associated with functions of the type <e, t> because this is the function of an entity toward a truth value: e → t. Transitive verbs and prepositions (represented by predicates with two arguments) are interpreted as functions of the type: <e,<e, t>>, because it consists of a function of an entity toward a function. Finally, nominal, verbal and adjectival groups are of the type <e, t>. Their domains can be divided into subsets of various types depending on the case: D_t, D_e, $D_{<e, t>}$, etc. Here is an example of a typed lambda formula: $\lambda P_{<e\ t>}\lambda x_e P(x)$.

Because indicating types weigh the terms down, they are often assumed without being given explicitly.

2.2.3.3. *The semantics of lambda expressions*

Lambda expressions can often be interpreted in two different ways:

a) $\lambda\alpha[\varphi]$ the smallest function that connects α and φ.

b) $\lambda\alpha[\varphi]$ the function that associates α with 1 if φ, or with 0 otherwise.

Interpretation *a* is used when φ does not have a truth value and interpretation *b* is used in the opposite case. When φ is associated with a truth value, then $\lambda\alpha[\varphi]$ is a characteristic function of a set. Therefore, lambda expressions can be considered to be a notational equivalent of sets as in the example:

$\{x \in D \mid \text{to run}(x) = 1\}$ $\lambda x.\text{to run}(x)$

This is a function that attributes the truth value 1 to *x* if the predicate to run (x) is true and 0 if the predicate is false.

2.2.3.4. *Conversions*

There are three rules that make it possible to convert or reduce the lambda expressions, two of which are relevant for this presentation: α-conversions and β-conversions.

Lambda expressions that differ only in names of variables are freely interchangeable. In other words, α-conversion stipulates that the names of variables have no importance in the expressions. The following expressions are therefore totally identical:

$(\lambda x.x) \equiv (\lambda y.y) \equiv (\lambda z.z)$

$(\lambda x\lambda y.F(x)(y)) \equiv (\lambda z\lambda x.F(z)(x)) \equiv (\lambda w\lambda z.F(w)(z))$

These expressions are sometimes called α-equivalents.

β-conversion (β-reduction) is used to represent the meaning of the components of the propositions. The lambda expressions are converted by this operation to obtain a representation of the entire proposition in first-order logic, which can be used by theorem-proving algorithms. The application of the β-conversion to a formula like: $\lambda x.P(x)@(a)$ consists of replacing all of the occurrences of *x* by *a* which gives: P(a). Consider the following examples:

– λx.eats (Paul, x)@(apple) $\rightarrow$ eats(Paul, apple)

– λxλy.searches(x, y)@(John)(Mary)[15] → searches(John, Mary)

– λx.strong(x) Λ intelligent(x)@(John) → strong(John) Λ intelligent(John)

This operation also applies to predicates that make it possible to define relations as in these two examples:

– λP.P(a)@(Q) → Q(a)

– λyλz. want(y, z)@(John)(λx. travel(x)) → want(John, λx.to travel(x))

2.2.3.5. *The use of lambda calculus in natural language analysis*

According to the principle of compositionality, the meaning of an utterance depends on the meaning of its components as well as its syntactic structure. Certain semantic approaches, like that of Montague, advocate for the direct matching between syntax and semantics by associating a semantic rule with each syntactic rule. The semantic representation of basic syntactic categories will be examined first, followed by the representation of two complete sentences.

Proper nouns are type *e*. Some consider proper nouns to be sets of properties, a set of sets, and attribute them the type: (e → t) → t. The lambda expression corresponding to a proper noun like *John* is: λP.P@(John). **Common nouns** like *girl*, *village* and *chair* are considered to be properties or sets and therefore of the type (e → t). Typically, the lambda expression corresponding to a common noun like *shark* is in the form: λx.shark(x) expressed as the property of *x* such as *x* is a shark.

Determinants apply to common nouns and convey the higher level structure of the nominal group. Processing determinants occurs in combination with universal and existential quantifiers as in the following cases:

– **A/An:** to represent this determinant, we can use the expression: λQ.∃x(P(x) Λ Q(x)). In this expression, the capital *Q* is a *complex variable*. It marks the place of complex information that is missing from the current stage of the processing. Some researchers prefer to note these variables with the operator @, as in: λP.∃x(P(x) Λ **Q@x**. Processing a nominal group like *a house* occurs according to the tree in Figure 2.16.

15 β-conversion applies to a single variable at a time and from right to left. In this case, the conversion occurs in two steps. The first has the result: λx.searches(x, Mary)@(John) and the second gives the final result: searches(John, Mary).

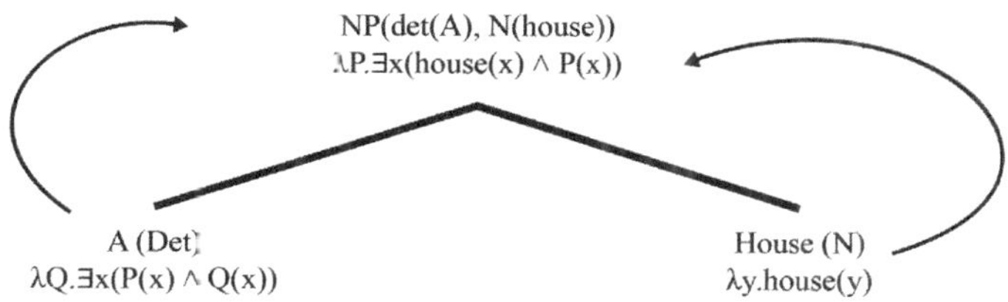

Figure 2.16. *Representation of a nominal group with the determinant a*

Complex variables also play a fundamental role in the representation of the other determinants:

– Each/All: the corresponding lambda expression is in the form: $\lambda P.\lambda Q.\forall x(P(x) \rightarrow Q(x))$.

– For example, all the students: $\lambda P.\forall x(student(x) \rightarrow P(x))$.

– Some: the lambda expression of this quantifier is: $\lambda P \lambda Q.\exists x(P(x) \wedge Q(x))$.

– For example, some journalists: $\lambda P.\exists x(journalist(x) \wedge P(x))$.

– None or no: this uses the expression: $\lambda P \lambda Q.\neg\exists x(P(x) \wedge Q(x))$.

– For example, no children: $\lambda P.\neg\exists x(child(x) \wedge P(x))$.

There are two opposing perspectives regarding the type of the **adjective** (see [TRU 04] for a more detailed discussion). On the one hand, some believe that the adjective denotes properties of the type (e, t). As shown in [PAR 87, ZAM 00], in constructions like [2.14] where an adjective can be coordinated with an indefinite determinative phrase, the adjective has a similar status:

I consider the president intelligent and an authority in his country. [2.14]

Je considère le président intelligent et une autorité dans son pays.

On the other hand, the standard perspective of categorial grammar that is defended by [LEW 70, KAM 75] among others considers that the adjective has an attributive nature and therefore plays the role of a predicate or a functor that takes a nominal element as an argument. This nominal element being of the type (e, t), they conclude that the type of adjective must be ((e, t) → (e, t)). Thus, the representation of an adjective like green becomes: $\lambda Px.(P(x) \wedge green(x))$.

Cases where an adjective does not modify a noun, as in [2.15], are worth mentioning. According to [KAM 75], these cases can be processed by using an implicit hyperonym. Thus, sentence [2.15] becomes Michael is a smart /boy/man/etc.

Michael is a smart. [2.15]

Verbs play the role of a logical predicate and can take zero or several arguments depending on whether they are transitive or not. The case of intransitive verbs is the simplest: they are type $(e \rightarrow t)$ and are associated with expressions like: $\lambda x.V(x)$. For example, the verb to run is associated with the expression: $\lambda x.\text{to run}(x)$. Processing transitive verbs logically is more complex. A naive solution for the processing could consist of using formulas like the one for the verb to eat: $\lambda x \lambda y.\ eat(x, y)$. This simplistic solution is obviously not correct, given that the object of the verb is none other than a nominal group, which is in turn a structure composed of a predicate with its arguments. Therefore, the expression for a verb like to eat is: $\lambda X \lambda z.(X@\lambda x.eat(z, x))$.

To illustrate the use of lambda calculus in the analysis process, consider a simple phrase without a quantifier such as: John looks at a flower. Analyzed syntactically, it provides the dependency tree given in Figure 2.17.

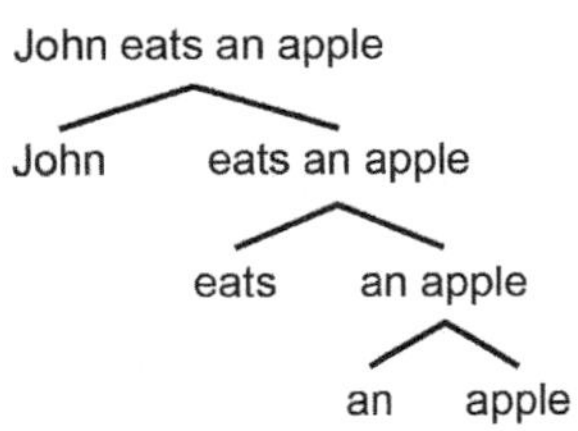

[S[NP John][VP[eats NP[an apple]]]]

Figure 2.17. *The syntactic dependencies of the sentence: John looks at a flower*

The semantic relations needed are:

John	$\lambda P.P@(John)$
eats	$\lambda X \lambda z.(X@\lambda x.eat(z, x))$
apple	$\lambda x.apple(x)$
An	$\lambda P.\exists x(Q(x) \wedge P(x))$

The analysis is obtained by a successive application of β-reductions (provided in Figure 2.18).

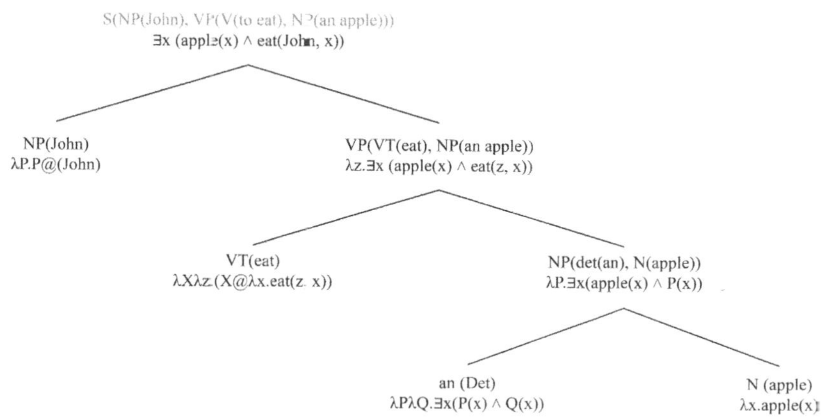

Figure 2.18. *Analysis of a simple sentence with a transitive verb*

2.2.4. *Other types of logic*

The name "first-order logic" suggests that there are higher level logics. One of the specificities of first-order logic compared to predicate logic is to allow quantification on objects. Higher level logics allow quantification not only on objects but also on properties or relations (predicates). For example, second-order logic allows quantification on predicates and functions as in the following case:

$$\forall x\ \forall y\ [(x=y) \leftrightarrow (\forall color\ color(x) \leftrightarrow color(y))]$$

This formula means that if x and y are equal, then they must be the same color. Third-order logic, in turn, makes it possible to quantify on the predicates of predicates, among others.

Intended for the formalization of modal information, so-called modal logic has various forms, including classic modal logic, epistemic logic (regarding knowledge), deontic logic (moral), temporal logics, etc. As an example, classic modal logic has four modalities: necessary, contingent, possible and impossible (see [BLA 01] for a formal introduction). This logic is particularly useful for modeling both human–human and human–machine dialogues [CAE 95, VIL 04].

Dynamic logics show how the representation of propositions evolves with the introduction of new information during the discourse.

3

The Sphere of Discourse and Text

3.1. Discourse analysis and pragmatics

3.1.1. *Fundamental concepts*

As noted by [VAN 85], the practice of analyzing speech, literature or simply language goes back more than 2000 years. The first works had a marked normative dimension and were located in the domain of rhetoric. They were intended to formulate rules for planning, organizing and delivering spoken communications in a legal or political context.

In the context of modern works in linguistics, notably since Zellig Harris [HAR 52], it is generally accepted that the sentence cannot be the maximal unit of linguistics studies. Consequently, many linguistic works consider that linguistic productions are made up of a set of interconnected utterances whose interpretation depends on the situation of communication. Some use the term *discourse* to designate such a set. Unfortunately, this term is one of the most polysemic terms that exists. The most precise definitions that have been proposed for it are those that have been formulated *negatively* in opposition to other linguistic entities.

One of the reasons behind this divergence is the multitude of movements and disciplines that have an interest in extra-sentential phenomena: functional linguistics, cognitive linguistics, sociolinguistics, textual linguistics, discourse analysis, etc. The common point between all of these approaches is the rejection of the Chomskyan idea expressed in his Standard Theory [CHO 57] according to which the sentence is the maximal linguistic unit.

Before addressing the key concepts in the domain of discourse analysis, it is pertinent to review the terminology.

3.1.1.1. *Discourse versus speech*

Like the French linguist Gustave Guillaume, some consider that discourse, being the language implemented or the language used by the speaking subject, can be considered to be a synonym of speech (see [DUB 71]). Linguists in this current prefer the opposition language/discourse to the Saussurian dichotomy language/speech because the term speech refers exclusively to spoken language.

3.1.1.2. *Discourse versus sentence*

Between the sentence and discourse, the boundaries seem relatively clear because it is generally accepted that discourse is a supra-sentential entity. However, as noted in [BEN 66], the nature of relationships between sentences is completely different from relationships between sub-sentential units like phonemes, morphemes or syntactic units. This makes works about discourse a specific domain within linguistics.

3.1.1.3. *Discourse versus narrative text*

According to the French linguist Emile Benveniste, the situation of utterance production is the difference between discourse and narrative text. Normally constituted discourse refers to the space-time of the utterance production and the *here* and the *now* refer to the place and time of the utterance production, respectively (see [LAC 97] for a detailed discussion). The narrative text, generally a set of written utterances, is characterized by its detachment from the situation of utterance production with a total absence of the personal pronoun *I*. Thus, in a narrative text, the third person is most common and the past tenses are used, such as the simple past and past anterior.

3.1.1.4. *Discourse versus text*

The most delicate distinction concerns the differentiation between discourse and text. Indeed, several currents differ on this subject. In order to give an idea of these divergences, Table 3.1 summarizes the main perspectives.

All of these points of view have advantages and disadvantages. From an automatic processing perspective, it seems that the point of view presented in [GAR 03] is particularly pertinent. Like Kamp and Reyle [KAM 93], he considers that discourse is a sequence of connected sentences and that this sequence has its own semantic representation. Therefore, the interpretation of discourse is considered to be an incremental process, where the semantic representation of a sentence must be connected to that of preceding sentences. A text, considered as a sequence of discourses, is defined according to the nature of the connections that exist between the discourses that compose it. This led Joaquín Garrido [GAR 03] to consider that the nature of these connections is what determines the genre of the text.

	Text	Discourse
1	The text is composed of sentences that have the property of being grammatically cohesive. Text analysis focuses on the cohesion.	The discourse is made of utterances, sentences used in a specific context with a given communicative goal, which have the property of being coherent. Discourse analysis focuses on coherence.
2	The text is a physical object without meaning.	The discourse is a process whose meaning emerges from the interaction between the reader and the text.
3	Texts are written.	Discourse is from the domain of oral communication.
4	Text analysis is part of discourse analysis.	Discourse analysis pertains to all extra-sentential phenomena including those observed in texts.

Table 3.1. *Some perspectives on the difference between text and discourse*

Some, like Richard Party [PAT 93], believe that discourse can be defined as an entity that has its own characteristics like uniqueness, continuity, intentionality and topicality. Uniqueness allows discourse to be perceived as an independent whole. Continuity, guaranteed by the presence of particular elements like discursive connectors, emerges from extra-sentential semantic phenomena. Intentionality focuses on the fact that discourse is above all an act of communication through which the locutor interacts with one or several interlocutors. Topicality pertains to the fact that a discourse must concern a specific subject that is identifiable by the locutor.

Generally, the actualization of discourse involves several levels of structuring at the same time. Such complexity gives rise to individual variations that are not always possible to predict, hence the difficulty of formally modeling discourse. Similarly, this multi-level nature opens the door to fairly varied interdisciplinary works with the objective of clarifying its different facets: psychological, social, ethnological, etc.

3.1.2. *Utterance production*

Benveniste considers utterance production to be a language mobilization process for the locutor. In other words, he considers it as an individual act of using language in a communication context whose final product is an utterance; the utterance being a linguistic entity directly related to the sentence. The sentence is a non-actualized language entity and therefore completely independent of all utterance production situations and their interpretive implications. The utterance, well-anchored in the situation of utterance production, finds its possible interpretations restricted and sometimes reduced to a single possible interpretation. In a certain way, the utterance can be considered to be a sentence anchored in a situation of utterance production.

3.1.2.1. *Deictics (shifters)*

The anchoring of utterances in a discourse is ensured by a particular category of words: deictics or shifters. This is a class of words that do not have a proper reference in language and only receive a meaning when they are included in a message. Linguists like Benveniste and Jakobson studied how facts of language relate to the context of utterance production. Deictics include a relatively large set of grammatical categories such as demonstratives, adverbs of time and space, personal pronouns and articles.

Deictics can refer to several aspects of the elocution context like:

– The speaking subject or the modalization. For example, *I told it to him*. Other categories of words like adjectives and possessive pronouns like *my*, *mine* or *Papa* can also be associated with this category.

– The time of the utterance production through words like *now*, *today*, *tomorrow*, *in a week* at the moment of the utterance. For example: *The weather is nice today*. Jakobson considered that because of their capacity to signal an anterior or posterior event at the time of the utterance production of the message, verbal tenses should also be included in this category.

– The space in which the utterance is produced through words like *here*, *beside* and *over there*. For example, *you can set that here*.

3.1.2.2. *Participants in a communicative event*

In the classic models of communication, such as those presented by [DE 16, SHA 48], two participants are involved in the communication. One is charged with encoding or producing and the other is charged with decoding or understanding. In the context of utterance production, things are not so simple. Such analyses have identified four roles that can be filled by one or more people according to the case.

– The locutor is the one that physiologically emits the (oral) message, or the writer or author of written messages. It is worth mentioning that the subject of the utterance (grammatical subject) does not necessarily correspond to the subject of the utterance production (locutor). For example, in the sentence *Paul eats an apple*, Paul is the subject of the utterance whereas the subject of the utterance production or the locutor is the person who pronounced this utterance.

– The utterer is at the center of the communicative act because this is the participant who assumes the responsibility for the content of the message. The locutor and the utterer can be the same person or a different person.

– The allocutor is the participant to whom the message is materially intended. This is sometimes also called the receiver or the recipient.

– The addressee is the participant for whom the message is formulated. Note that, unlike the utterer, which is always singular because even in the case of *we* there is always an *I* behind it, the addressee can be plural, because a theoretically unlimited number of people can be addressed at the same time.

An extreme case of a communicative act is a monologue, when a person of sound mind talks to himself. In this situation, all of the participants are the same person. This observation is only valid in true monologues that can be observed in real life. In false monologues, which are relatively frequent in artistic works like operas, novels and plays, the monologue is used to reveal the feelings or thoughts of a character to the audience. In that case, things are quite different. First, the locutor and the utterer are no longer the same person: the locutor is the actor and the utterer is the character being played by the actor, and sometimes the author of the text performed. Second, even if the person is talking to himself, the true addressee of the message is the audience.

Another case that should be mentioned is quotes or indirect or reported speech. In this case, the locutor borrows the comment of another person, be this in a literal way or not. For example, in the case of [3.1], the utterance is pronounced by the professor, who is in fact the locutor, but the content is visibly not assumed by the professor. Rather, it is attributed to Michael, who is considered the utterer.

> The history professor said to his students, "Your classmate Michael wrote, 'World War I took place in the 19th century.'" [3.1]

Note that the distinction between the utterer and the locutor is not always clear. Indeed, between a total assumption and a complete detachment from the message's content, there are nuances that can be configured by the locutor, notably through the use of the conditional or modal adverbs like *most likely*, *maybe* and *probably*.

There are also many cases where the allocutor and the addressee do not correspond to the same participant. This case is especially common in journalistic interviews on television or in the written press. The interviewee, in responding to questions, speaks to the journalist, the interviewer, who is the allocutor, but the real addressee of the message is the audience who watches or reads the interview.

When a message involves several participants (voices) during its production and/or is intended for different audiences during its reception, this is called polyphony. Initially observed by Charles Bally and Mikhail Bakhtin, this phenomenon was taken up and developed by [DUC 87]. Note that Gérard Genette, the eminent French literary theorist, was also interested in similar cases in literature where it is possible to distinguish between the author, character and locutor. These literary studies were conducted under the label of narratology [GEN 07].

Finally, it is important to mention that the distinction presented here is not unanimously accepted by the community. For more details about the differences surrounding this issue, consult [RAB 10].

3.1.2.3. *Time and space adverbials*

In a discourse, a chronology makes it possible to situate events in relation to the moment of utterance production. It can be marked through a multitude of grammatical means such as articulators like *first of all*, *in the first place*, *then*, *having said that* and *finally*; conjunctions like *when*, *once* and *after*; time adverbs like *today*, *yesterday* and *tomorrow*; and nouns indicating time like *Monday*, *night* and *evening*. The use of verbal tenses like the present, future and past is the simplest way to express the chronological order of events.

The spatial situation of events is generally indicated by space adverbials like *here*, *there*, *elsewhere* and *around*. The adverbial *here* can be analyzed as the place where the *I* commits the act of utterance production.

3.1.3. *Context, cotext and intertextuality*

The factors examined above are, without a doubt, elements that are essential to what can be called context. However, the concept of context can extend far beyond these factors. As stated in [GOR 09], the concept of context is the main reason for the divergences between the different approaches to discourse analysis.

For example, in the analysis approach to verbal interactions presented in [KER 96], the concept of context includes the following elements: a spatiotemporal framework, a goal and participants.

The spatiotemporal framework, or the place where the interaction occurs, directly affects the nature of the interaction in various ways. On the one hand, the physical nature of the place (public place, open, office, etc.) can restrict the subjects and the way in which they can be addressed. For example, people tend not to discuss intimate topics like family problems in public places. On the other hand, the social, religious or institutional role of a place implies particular subjects and even a certain level of language. For example, people tend to be more formal and avoid using vulgar words in places of worship or a courthouse.

As for the goal, there is a distinction between transactional interactions that have a specific goal, like purchasing a plane ticket or requesting information about a given subject, and relational interactions whose only aim is to maintain a social relation, like chatting with friends.

Several factors intervene when it comes to determining the role of the participants:

– The number: people do not speak in the same way as in a one-on-one conversation when they are in front of a large audience like at a conference or in a televised interview.

– The individual properties: factors like age, gender and social status considerably affect the linguistic register to adopt.

– Familiarity: sometimes called horizontal distance, this factor is related to the change in linguistic behavior depending on the degree of familiarity between the interlocutors. For example, people tend to be more formal with strangers than with people they know well or are close with like friends, work colleagues or family members.

– The cultural context: cultural factors can play a very important role in the choice of the form of discourse. The status of elderly people and women is not the same in all societies. Similarly, some societies are very hierarchized, especially Eastern societies like India and Japan, where very polished language is used when addressing someone of a social rank that is considered superior.

The concept of cotext concerns textual context or the constraints that one part of a discourse exercises on the subsequent parts. Thus, aside from the initial part of a discourse, the interpretation of any part of a discourse must take the previous parts into consideration. To a certain extent, the cotext is a sub-type of the context. Consider passage [3.2], where the interpretation of the pronoun *we* as equal to (I + Samir + his friend Steven) cannot occur without using information conveyed by the previous part of the text.

> I spent a nice day in Normandy with my friend Samir and his friend Steven. To return to Paris, we took the train at 8 p.m. [3.2]

The concept of intertextuality or transtextuality[1] is used in situations where another conversation, discourse or text is referenced. This concept has been the topic of several studies, especially in literature where they focus on the relationships that exist between a text and pre-existing texts to which they refer in order to oppose or support them. The intertext of a given text is the set of writings to which the text is related, explicitly or not.

3.1.4. *Information structure in discourse*

3.1.4.1. *Topic and comment*

The topic is what the text/discourse is talking about. It is part of the information that is generally accepted or known by the interlocutors, either because it is provided by the text/discourse prior or because it is shared by the interlocutors. This information can concern generalities about the world like *the sky is blue*, *gold is expensive* and *the moon rotates around the Earth*, or individual information shared by the members of a given group such as members of the same family, work colleagues or members of a sports team.

The topic depends on the order of the sentence. Thus, two sentences that are considered semantically equivalent can have different topics if their elements are ordered differently. For example, in sentences [3.3], the topic in (a) is the agent *the cat* which in (b) becomes the patient *the mouse*:

a) The cat chased the mouse. Topic: the cat.

b) The mouse was chased by the cat. Topic: the mouse. [3.3]

In the case of sentences with an indefinite subject, the semantically empty subject has only a purely formal role in the sentence and consequently does not have a topical role (see sentences [3.4]):

– It is raining. Topic: the whole proposition. [3.4]

– There is bread. Topic: bread.

The comment or rheme concerns what the text says about the topic. It consists of new information that the text provides about the topic (see [3.5]).

– Saturday morning, a car hit a small cat. Comment: a small cat

– It was Adeline who did all the work. Comment: Adeline. [3.5]

1 Again, the terminology is not unanimously accepted here.

3.1.4.2. *Topical progression*

Because a coherent text does not necessarily concern only one subject or topic, it is essential to study the mechanisms of topical progression. Topical progression pertains to the relationship between the topic of a sentence with the topic and comment of previous sentences. Four forms of topical progression are often cited in the literature: constant topic progression, linear topic progression, derived topic progression and inserted topic progression.

In the case of constant topic progression, the topic is shared by several successive sentences. This type of progression is particularly useful when describing an object or a character by mentioning its different properties and actions. This form of succession is commonly observed in narrative texts. Very often, the grammatical topic is confused with the grammatical subject. The use of anaphora is the natural method for this kind of repetition. Schematically, this kind of progression can be presented as shown in Figure 3.1.

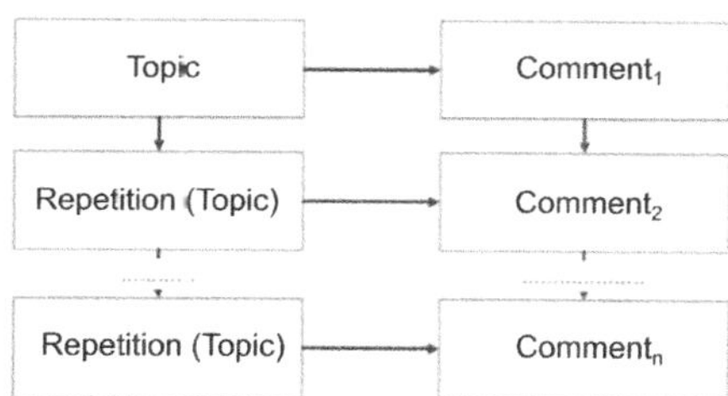

Figure 3.1. *Diagram of constant topic progression*

To make this type of progression clearer, examine text [3.6] where the topic Fabrice is repeated in the sentences either by the personal subject pronoun he or by the possessive adjective his:

> Fabrice is a graduate of the Faculty of Pharmacy in Boston. Two years ago, he went to do a six-month internship in Copenhagen at a famous Danish company. There, he learned many things about the pharmaceutical industry. Upon his return to the US, he managed to get a good job in New York. [3.6]

In linear topic progression, each sentence activates another one. This occurs through a comment/topic exchange: the comment of one sentence becomes the topic of the following one and so on. A diagram of this kind of progression is provided in Figure 3.2.

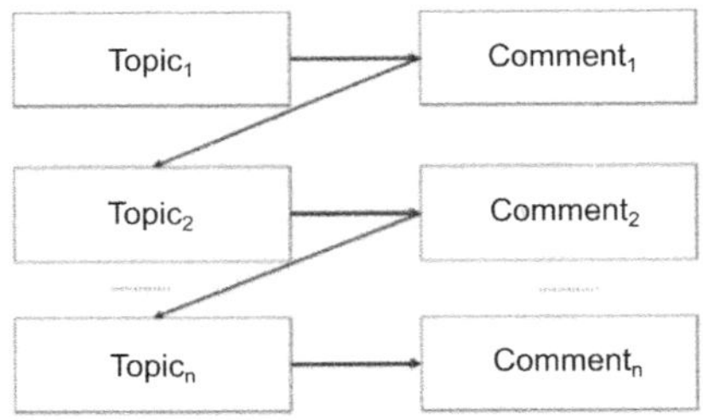

Figure 3.2. *Diagram of linear topic progression*

> Ashley listened attentively to the words of Marilyn, her best friend. These words conjured up the bitter memory of her break-up with her ex-fiancé Mark. It was this break-up that pushed her to attempt suicide and then to shun everything to do with men. [3.7]

With derived topic progression, there is a main topic presented first and then derived (less central) topics follow. Often, the relationship between the main topic and the derived topics is meronymic (whole/part) or hyponymic (type/sub-type). See Figure 3.3 *for the diagram of this progression.*

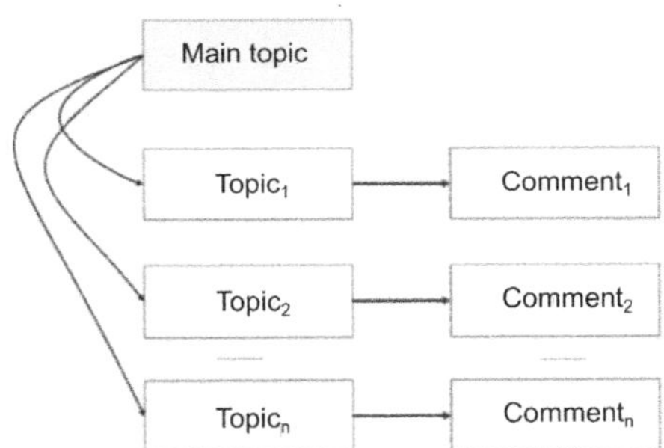

Figure 3.3. *Diagram of derived topic progression*

In passage [3.8], the relationship between the topic of the first utterance, Venice, and the topics of the subsequent utterances is meronymic. Canals, castles and gondolas are all components of the city of Venice:

> Venice is by far the most beautiful city I have ever visited. Its canals made me forget about the stress of everyday life. Its marble castles made me think of the adventures in One Thousand and One Nights. The gondolas reminded me of the romantic period of my engagement. [3.8]

Finally, with inserted topic progression, it is sometimes possible to insert a local topic related to the main topic of the text within a constant topic progression. This results in the diagram provided in Figure 3.4.

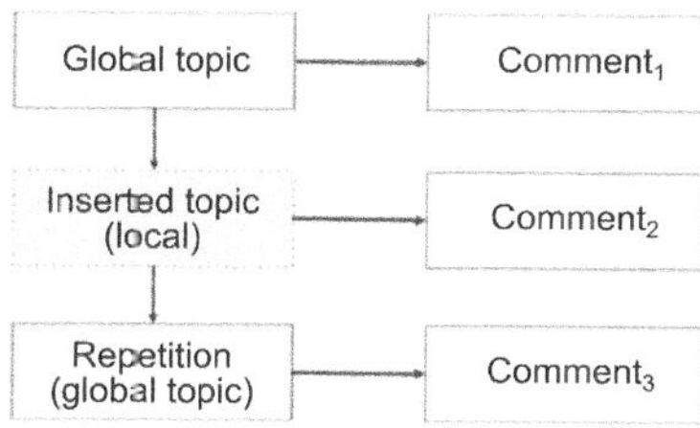

Figure 3.4. *Diagram of inserted topic progression*

Consider example [3.9] to illustrate this kind of progression. In this text, the discursive topic is *Frank's car*. A new local topic is added in the following sentence, which is Frank repeated by the anaphor *this man*. This does not prevent new comments on the discursive topic from being added to the third utterance:

> Frank's car is really superb. This man, who is passionate about mechanics and everything related to the automobile industry, takes great care with what he owns. Even though it was manufactured in the 1960s, the car is in excellent shape. [3.9]

It should be noted that, naturally, authors do not hesitate to combine the types of topical progression to avoid monotony in their texts even if a particular form of progression generally prevails depending on the genre of the text. This wealth of types of topical progression makes the automatic detection of topics addressed by a text particularly delicate work.

3.1.4.3. *The implicit, the presupposed and the assumed*

The interpretation of an utterance does not only occur through the elements that are explicitly present. Often, it requires inferences about new elements (not given in the sentence) from the given elements. This information is called the implicit. The study and formalization of these interpretive mechanisms of this kind of utterance attracted the interest of several researchers (see, for example, [KEE 71, KEM 75, DUC 69]).

For example, from an utterance as simple as *Close the door*, we can infer the following information about this utterance:

a) The locutor has a certain authority over the allocutor, as indicated by the use of the imperative.

b) There is a door in a space nearby and it is open or at least that is what the locutor thinks.

c) The locutor thinks that the allocutor is capable of closing the door.

d) The locutor has a specific goal behind this request: he is cold, he wants to have a private conversation with someone in the room, he wants to start his break, etc.

In this example, the inferences *a*, *c* and *d* were made based on the utterance context whereas *b* can be inferred in all possible contexts because it depends on the verb *to close*. Some researchers prefer to talk about the implicit of the utterance production (the information in *a*, *c* and *d*) and the implicit of the utterance (the information in *b*). A classification on similar criteria consists of distinguishing two sub-types of the implicit: presupposition and the assumed.

A presupposition is information deduced from a lexical unit in the utterance. Such a unit could be an adverb like *again*, *already* or *always* that suggests an event prior to the moment of utterance production. It can also consist of an adjective, a *verb*, a superlative, etc. Consider the examples in Table 3.2.

Sentence	Presupposition(s)
Paul still smokes.	Paul used to smoke before.
Michael does not smoke in the house anymore.	Michael used to smoke in the house before.
Mary was not feeling well last night.	She was feeling well before that time.
"Star Wars" is Jeff's favorite film.	Jeff likes other films.
Cycling is Joe's favorite sport.	Joe likes other sports.
William is smart but Arnold is strong.	William is not strong.

Table 3.2. *Examples of sentences with their presuppositions*

The main difference between the assertion and the presupposition is that only the assertion is the focus, which is why it alone is affected by negation. For example, in *The volcano in Paris is erupting*, there are two distinct items: 1) there is a volcano in Paris (presupposition) and 2) this volcano is erupting (assertion). In the negative form *the volcano in Paris is not erupting*, it is the assertion that is refuted, the fact that the volcano is erupting. The presupposition remains valid, although this does not necessarily mean that it is true.

On the other hand, the assumed is information that the locutor implies without giving it and without the presence of a linguistic cue in the utterance that would make it possible to deduce it. Consider utterance [3.10] that can receive one or more of these interpretations: let's go to another location, close the window, turn down the volume on the television, I cannot hear you clearly.

It is very noisy here. [3.10]

Thus, the addressee of the message with an implicit is obliged to infer the information because, taken literally, it does not have any significance for him or at least it seems marginal. In other words, in this case, the allocutor is entirely responsible for the interpretation constructed. Consequently, it can be said that, unlike the presupposition, the locutor has the possibility of negating the assumed. In practice, because it does not directly engage the locutor, the implicit is a very good way to make an observation to someone or to make a potentially embarrassing request.

3.1.4.4. *Given versus new*

The dichotomy given/new concerns what is known in opposition to what is unknown. Traditionally, two types are distinguished by the use of syntactic criteria: indefinite articles are used to mark nominal groups that correspond to unknown entities and after their introduction into the discourse, these entities are referred to by definite articles or anaphora ([HAR 51] cited in [BRO 88]).

Today, there are several diverging points of view regarding the specification of given or unknown information [CHA 76, CHA 87, HAL 70]. This presentation will adopt that of [PRI 81] who considers the different aspects of the problem. Starting from the observation of the simple dichotomy of new/old to consider the complexity of the organization of the information in the discourse, Prince proposes a taxonomy that includes three classes: new, inferrable and evoked. She proposes a tree diagram of this taxonomy, which is shown in Figure 3.5.

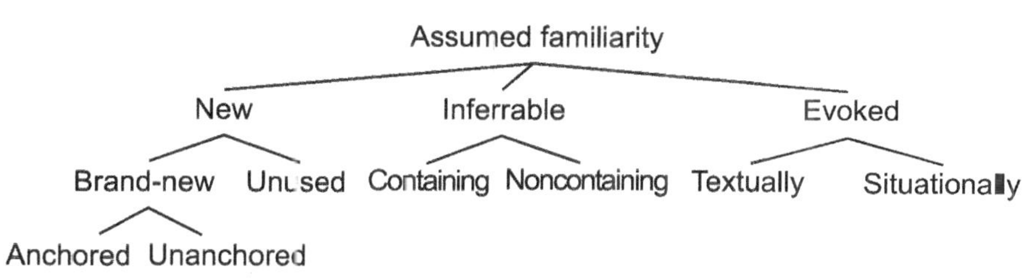

Figure 3.5. *Prince's taxonomy [PRI 81]*

The term *new* concerns new information introduced to the discourse. There are two sub-types: new information for the interlocutor and new information for the discourse. In the first case, a completely new entity is introduced for the interlocutor for which he must entirely create the mental representation. In the second case, where an entity that is already known to the locutor is introduced into the discourse, the interlocutor must update his representation of the discourse to include this new entity. Consider the utterances in [3.11]. Assuming that these utterances are found at the start of a discourse, the entities *the sea* and *Paul Broca*, newly introduced in the discourse, are assumed to be known by the interlocutor. On the other hand, the entities *a person* and *a store* are new to both the interlocutor and the discourse. It should also be noted that Prince distinguishes between two types of these entities: anchored entities and unanchored entities. Anchored entities including the ones such as *a person* are considered as such because they are syntactically related to a nominal group that is not new, *I*:

a) *The sea* is very rough today. [3.11]

b) *Paul Broca* was born in Gironde.

c) *A person* with whom *I* traveled told me she read your book.

d) I entered *a store* where the manager was very angry.

Evoked entities are both known by both interlocutors (shared or supposedly known knowledge between the two interlocutors) and already introduced into the discourse. Prince distinguishes two sub-types of evoked entities: textually evoked entities and situationally evoked entities. Textually evoked entities are entities that have already been mentioned in the text, often by means of third person pronouns or a nominal group with a definite article. Situationally evoked entities are entities that have been indicated by a person, time or space deictic and whose interpretation is anchored by the situation.

Contrary to evoked entities, inferred entities do not have an explicit antecedent. Their presence can be guessed using an element that serves as an indirect antecedent. As emphasized in [CHA 76], these entities, which are not exactly new, are not entirely given either, because they cannot be replaced by an anaphor. Thus, inferred entities can be considered to be an intermediary category between given and new. For example, in [3.12], the use of a definite article before *cotton wool* indicates that even if this entity is being introduced for the first time in the discourse, its presence in the first aid kit can be inferred logically. This means that *first aid kit* can be considered to be an indirect antecedent of cotton wool:

We checked the first aid kit. Everything was there but the cotton wool was missing. [3.12]

3.1.5. *Coherence*

Coherence is a property of the content of a text that assembles words and sentences in one or more connected discourses. It is this property that allows a text to be perceived as a unit equipped with a meaning that conforms to a certain vision of the world. From the point of view of its content, a text can be considered to be a set of concepts and relations [DE 81]. These concepts, viewed as cognitive content, can be activated or, inversely, inhibited depending on their relevance to the context and their relations to the other concepts. Consider the texts [3.13] and [3.14]:

> – The sonata, from the Italian *sonare* that means to perform with an instrument, is a musical composition intended for one or more instruments. In France, some particular forms of the sonata like the cyclic form and theme form have been used by composers like Hector Berlioz and Gabriel Fauré. [3.13]

> – The bicycle is an excellent mode of transportation in both urban and rural environments. It is also one of the most commonly practiced sporting activities in winter, especially in countries like Canada and Iceland. The increase in train accidents in the south-west of the country is causing panic among travelers. [3.14]

The utterances in [3.13] seem logically related, unlike the utterances in [3.13]. Therefore, it can be said that [3.14] is a coherent text whereas [3.14] is not.

The concept of coherence is closely related to the quantity of information shared between the locutor and the addressee. Depending on the contextual information shared between the locutor and the addressee, a discourse can be determined to be coherent or not. Consider the mini dialogue in [3.15]. This mini dialogue can only be perceived as coherent if the addressee of utterance B (A) knows that John plays a main role in the transportation process. For example, as the driver of the truck transporting the merchandise:

– A Did you transport the merchandise to the store?

– B John went to see his father at the hospital. [3.15]

Among the most commonly observed relationships between the utterances of a coherent discourse are relations of causality and relations that express a chronological order.

The relation of causality concerns the way in which a situation or an event affects the conditions of realization of another situation or event. Consider utterance

[3.16] where there is a logical cause and effect relation between (cause: driving with concentration) and (effect: Ralph's victory):

Ralph drove with such concentration that he won the race. [3.16]

When the information is conveyed by several sentences, these kinds of relations can be marked explicitly by the use of discursive connectives that indicate deductions such as *therefore, then, because of this, that is why* and *consequently* (see example [3.17]):

Ralph drove with a great deal of concentration. That is why he won the race. [3.17]

Note that the semantic nature of the relation between the two propositions of cause and effect can vary considerably. For example, this relation can take the form of:

– Logical consequence: if John was on a trip the day of the crime then we can say that it was not him that killed Mary.

– The ultimate objective of an action: Paul is working hard at school to become an engineer.

– The end and the means: by calling on his colleague, the technician was able to repair the machine.

Relations of chronological order concern the order of events. Sometimes, this order is hidden behind a causality relation [3.18a] but often it consists of a series of actions or events [3.18b]:

a) The soldier was shot in his left leg and then he fell to the ground. (chronology and causality).

b) He called the director and then he went to dine in the restaurant on the corner. (simple succession). [3.18]

3.1.6. *Cohesion*

Traditionally associated with the works of [HAL 64, HAS 68], the term cohesion can be defined as the set of means necessary for the construction of relations that transcend the level of grammar within a discourse [HAL 94] (see [MAR 01] for a detailed representation). Unlike coherence, cohesion is a property that concerns the structure of a text, notably its linguistic form, not its content. It is an objectively observable dimension because it is based on concrete elements. Cohesion can be seen as a necessary but not sufficient condition for coherence. In other words, on its own, cohesion is insufficient to interpret a discourse. Its role is limited to improving comprehension.

3.1.6.1. *Lexical cohesion*

To repeat a word from a previous sentence in a new sentence, anaphora seems the most natural means. However, two other important processes can be cited: repetition and the use of a lexical chain.

The simple repetition of a lexical unit several times is shown in [3.19]. This kind of repetition is often used in literature in order to emphasize something or to provide a sense of monotony or weight.

> *Sam* is looking at the flowers. *Sam* is walking in the garden. *Sam* is going to school. [3.19]

A lexical chain consists of one word with lexical substitutes that are distributed throughout a given text. The relationship between the lexical units in a chain can be synonymy, antonymy or belonging to the same lexical field. Consider fragments [3.20]:

> The *dog* is an incredible animal. *Man's best friend* does not hesitate to risk its life to save its master. (Synonym)
>
> The *dog* is an incredible animal. Sometimes, this canine/mammal/ carnivore does not hesitate to take risks to save its master. (Hyponym/Hyperonym) [3.20]
>
> The *dog* is an incredible animal. Like the *cat*, it is a top choice for a pet for many people. (Semantic field: domestic animals)

3.1.6.2. *Anaphora*

Anaphora is a very important way to ensure a link between different discursive units in speech and in writing. However, its role is more central in speech than in writing due to the dialogic structure that implies an exchange between two interlocutors and therefore requires referencing previously mentioned fragments of discourse.

The definition of anaphora that is generally given is that anaphora is a mechanism that relates two linguistic units. The first unit is often pronominal (a personal or demonstrative pronoun) and is called the anaphor. The second unit is an anterior segment, typically a nominal group (see [DUB 71] for example). Consider utterance [3.21a]. In this utterance, the nominal group *Peter* is repeated by the anaphor *him*. Note that sometimes the order can change, which means that the anaphor can come before or after the nominal group [3.21b]. This is called cataphora:

> a) Peter, I see him often.
>
> b) I often see him, Peter. [3.21]

A more specific definition of anaphora was proposed in [KRA 00]. It is based on several criteria to interpret anaphora like contextual dependency, the type of antecedent, the type of relation between the anaphor and the antecedent, and the interval of interpretations authorized by the anaphora. In a dialogue, two types of antecedents can be distinguished. The immediate antecedent concerns cases where the anaphor and the antecedent occur in the same conversational turn, as in utterance [3.21a]. In the case of a distant antecedent, the anaphor and the antecedent occur in two conversational turns that are potentially spoken by two different locutors. As an example, consider the segment in [3.22] extracted from a dialogue in a hotel reservation corpus collected at the CLIPS-IMAG laboratory [HOL 97]. In this extract, *the anaphor it and its antecedent* occur in two conversational turns of two different locutors, the customer C and the hotel receptionist H, respectively. Note the formal ambiguity of the attachment of the anaphor because in the utterance H, there are several nominal groups that could be candidates: a room, a person, a shower:

H = So, I have a room for one person with a shower and WC on the fourth floor looking out on the garden for 380 dollars, breakfast included. [3.22]

C = Very good, I'll take it.

There are even more complex cases where the anaphor returns throughout several segments of the discourse. In these cases, the anaphora is very difficult to detect automatically because it requires a very large contextual window that often contains several ambiguities to be considered.

Several linguistic factors make it possible to find the referent of an anaphor. Some are decisive and therefore impossible to violate while others are facultative but they allow the search field to be limited.

– The number, person and gender: obligatory, these distinctive features can direct the search (see examples [3.23]):

John visited his friend and his cousins. *He* gave him a gift.	The feature *singular* of the anaphor *him* makes it possible to decide that *his friend* is the referent of the anaphor.	
We went to Dallas with my brother last year. *He* really appreciated *your* presence.	Here, *we* is equal to *me* and *you* plus *him* eventually, the anaphor *he* refers to *my brother* and *your* refers to the second person mentioned in *we*.	[3.23]
Paul and Mary graduated this year. *She* had the best grades at her school.	The feature *feminine* makes it possible to attribute the reference *Mary* to the anaphor *she* because the other candidate, Paul, has the feature *masculine*.	

– Novelty: from a cognitive point of view, in case of a multitude of possibilities for the resolution, the most recent element, being the most prominent in the memory, has the advantage. Consider example [3.24]:

Randy was playing table tennis with Will. They met Barry in the room. *He* was very happy that day.	Naturally, *he* refers to Barry, the most recent candidate, and not Will or Randy.	[3.24]

– The topical role: an agent in a sentence has more chance of remaining in a second sentence. Examine examples [3.25]:

Kati saw Amanda with her newspaper in the garden. She said hello to her.	Kati, the agent of the first sentence, is logically the antecedent of the anaphor *she*.	[3.25]
Kara was seen in the garden by Nicole. She said hello to her.	Here, logically she refers to Nicole, the agent of the first sentence, even if Kara was the focus.	

– Repetition: the repetition of an anaphor throughout a text makes it possible to identify the referent in all occurrences of this anaphor. Examine text fragment [3.26] where the anaphor *she* is repeated several times, making it possible to identify that the referent is Cynthia and making it unlikely that Martha is the referent of the utterance: she went...

> Cynthia is a nurse. She likes her job a lot because she loves helping people. She works with her friend Martha in the same hospital. During the holidays last year, she went to Morocco with her husband. [3.26]

– Reciprocity or parallelism: sometimes the agent of a sentence becomes the patient or the recipient in another sentence and its patient or recipient becomes the agent in the other sentence. This logical reciprocity allows the referent of the anaphor to be deduced. For example, in passage [3.27], the agent in the first utterance, John, becomes the recipient in the second utterance, while the recipient of the first utterance becomes the agent in the second utterance:

> John gave a bouquet of flowers to Mary, Maurice's sister. Mary gave him a pen (him = John). [3.27]

– The semantics of the verb: sometimes the semantic relations between the arguments of certain verbs imply relations between the actants of the verbs, as in [3.28] where the agent of the verb *to send* is the recipient of the verb *to receive* and vice versa. This helps deduce that the referent of the anaphor *he*, the agent of the verb to receive, is Max:

Fred sent a postcard to his work colleague Max. *He* was very happy when he received it. (He = Max) [3.28]

3.1.7. *Ellipses*

Ellipses consist of omitting a certain number of elements from an utterance without affecting its intelligibility. The omission creates a puzzle effect that allows the auditor to find the omitted elements and complete the information. Like anaphora, ellipses are a linguistic phenomenon common in both spoken and written languages, but they play a more important role in spoken language, especially in responses to certain questions. In general, an ellipsis is an important way to avoid redundancies and consequently make the conversation simpler and more spontaneous. There are two types of ellipses: situational ellipses and grammatical ellipses.

Situational ellipses include a set of ellipses whose interpretation depends directly on the situation of elocution. As shown in the previous paragraphs, this situation can be the history of a dialogue, the physical context in which the conversation occurs, general knowledge about the world, etc. Consider the mini dialogue [3.29] where there is a situational ellipsis. In this example, note the double ellipsis in the response: the removal of the request formula *I would like* and the word *room*.

H: Would you like a single room or a double room?

C: A single, please. [3.29]

Grammatical ellipses consist of omitting words that syntactic knowledge of the language can make it possible to infer. The most commonly studied form of grammatical ellipsis is the verbal ellipsis [HAR 97]. In this kind of ellipsis, the verbal group is removed in contexts where it is considered inferrable, as in utterance [3.30]. In this utterance, the verb in the second proposition is removed, which suggests that it is the same verb as the one in the first proposition: eats.

Peter eats cherries and Paul strawberries. [3.30]

Mixed ellipses are possible in some contexts. To illustrate these ellipses, consider the example in [3.31]. In this exchange, we removed the segment I am which is easy to infer from the syntactic rule: subject + verb to be + qualifier, 'in agreement'. Note that syntax alone is sufficient to infer the verb *to be*. Syntax also played a direct role in inferring the subject, however the form of the subject (noun, pronoun) as well as the person (1st person singular, 2nd person plural, etc.)

requires discursive context. Thus, the final analysis of this ellipsis mobilizes both syntactic and contextual knowledge.

A: What do you think? [3.31]

B: Completely in agreement.

Some forms of ellipsis can be seen as particular cases of anaphora [KRA 00]. Ellipses are based on a strong connection to a previous part of the discourse, like anaphora. However, contrary to anaphora where a linguistic device is needed to refer to the previous part of the discourse, ellipses are characterized by the removal of elements shared with what has been said.

3.1.8. *Textual sequences*

Intuitively, when a reader reads a text, he/she is often able to label it as a text category: narration, description, etc. However, establishing such a categorization on linguistic bases is far from obvious. Several linguists have attempted to produce textual typologies like [WER 75, ADA 92, ADA 01].

Adam's classification is based on the notion of *sequence*, which he considers to be the constituent unit of the text. A sequence is composed of micro-propositions that are in turn composed of *n* (micro) propositions. Adam proposes five main types:

– Narrative sequences recount events that occur in a temporal or causal (chronological) link. Particularly common in novels, stories and narrative texts, they are also commonly used in journalism and sports reporting, some ads and accident statements. From a grammatical point of view, these sequences are marked by tenses like the imperfect, simple past and historical present, as well as adverbs and indications of time (now, tomorrow, two days later).

– Descriptive sequences essentially pertain to the description of the spatial arrangement of propositions. This concerns texts that indicate the moral and/or physical qualities of a person, thing, landscape, location, etc. These sequences are characterized grammatically by the use of the present or the imperfect, as well as adverbs and place names.

– Dialogical sequences, common in plays and some ancient philosophical texts, are also very frequently found in the different ways of interacting available today on the Internet like discussion forums and chat sites/software. The content of these sequences is distinguished by the management of speech acts as varied as promises, affirmations, threats, acknowledgements, etc.

– Argumentative sequences are centered on the defense of a position or particular point of view. These sequences follow a logical reasoning framework: linking one or more premises and a conclusion. They are common in different types of university projects like theses and dissertations, essays and some political discourses characterized by the adoption of a position in relation to a given project, person or idea. Linguistically, these sequences are associated with dialogue acts like *convince*, *persuade* and *cause to believe*.

– Explanatory sequences, like argumentative sequences, are frequently observed in university projects as well as handbooks and other forms of scientific books. They are intended to explain the how and why of a given phenomenon.

Note that Adam borrows the prototypical view of lexical semantics proposed in [KLE 90]. In other words, a text is no longer considered as belonging or not to a given category (discrete view) but rather as more or less typical or atypical, depending on its distance from the notional reference prototype of the category or discursive genre under consideration.

3.1.9. *Speech acts*

The theory of speech acts was developed by John Austin at the University of Oxford in the context of work on the philosophy of language that originated in the current of analytical philosophy [AUS 55]. Later, this theory was substantially developed by the works of John Searle at the University of California at Berkeley [SEA 69]. This theory is based on the distinction between two types of utterances: utterances that describe the world, called constatives, and utterances that modify it, called performatives (see examples [3.32]). The main property of constative acts is to receive a truth value. For example, in utterance [3.32a], the bush can be planted in the garden or not. On the other hand, truth values do not make sense in utterance [3.32b]. These cases include things like the sincerity of the person uttering the sentence, its capacity to realize its engagement, the failure or success of the act, etc. This concerns the condition of felicity rather than the condition of truth:

a) John plants a rose bush in the garden at his house.
(Constative) [3.32]

b) I promise to repay all my son's debts. (Performative)

Note that the boundary between the two types of acts is not completely clear-cut. For example, in [3.33], even though the utterer does not use a verb that explicitly indicates that he promises to buy the chocolate cookie, it is implicitly understood that this is a promise:

I will buy you a chocolate cookie tomorrow. [3.33]

The act committed by uttering a sentence with a performative value is called a speech act. There are three groups of speech acts: locutionary acts, illocutionary acts and perlocutionary acts.

3.1.9.1. *Locutionary acts*

These are acts that are accomplished as soon as a meaningful utterance is uttered orally or in writing, independent of the communicative role of this utterance. Consider utterance [3.33] whose phonetic, morphosyntactic and semantic formulation is a locutionary act.

3.1.9.2. *Illocutionary acts*

An illocutionary act is carried out on the world by producing the utterance. For example, a judge saying *I declare that court is in session* will effectively open the session. Austin describes five major types of illocutionary acts:

– Verdictive: acts pronounced in a judicial and legal context. For example, the following verbs express a verdictive act: *to declare guilty/innocent, to condemn, to decree, to acquit.*

– Exercitive: acts related to the exercise of authority generally in contexts as varied as military, administrative or familial contexts. The verbs *to command, to order, to rule, to dictate, to prescribe, to pardon* are examples of these acts.

– Permissive: acts that commit the locutor to carry out actions in the future. For example, verbs like: *to promise, to commit, to bet, to gamble, to guarantee.*

– Behabitive: acts related to behavior expressed by verbs like: *to thank, to bless, to praise, to criticize.*

– Expositive: acts concerning the relationship between the locutor and the information that he transmits. For example, *to affirm, to certify, to assure, to deny, to note, to notice, to observe, to discover.*

It should be noted that John Searle conducted a critical study of Austin's taxonomy and concluded by proposing his own. Like Austin's taxonomy, it has five groups: assertives (assertion, affirmation), directives (order, command, demand), permissives (promise, offer, invitation), expressives (thanks, praise) and declaratives (declaration of marriage, nomination for a prize).

3.1.9.3. *Perlocutionary acts*

These are acts that either have a voluntary effect or not on the allocutor. They are the consequences of an illocutionary act:

a) I will take you to the park next week. [3.34]

b) I declare that court is in session.

Beyond the locutory act related to the formulation of the utterances in [3.34] (both a and b) and the illocutionary act related to the commitment, utterance [3.34a] has a possible perlocutionary effect, which is to persuade the allocutor of the kindness of the locutor. If we assume that utterance [3.34b] was produced by a judge in a court, it has the perlocutionary effect of attracting the attention of the listeners to the judge, putting an end to the lateral conversations of the people present, etc.

Finally, it is useful to mention that the theory of speech acts saw important developments in the 1990s, notably in the works of Sperber and Wilson on Relevance Theory in the cognitive pragmatics movement [SPE 89] and the works of Albert Assaraf on Link Theory [ASS 93].

3.2. Computational approaches to discourse

The computational approaches to processing discourse are very divergent. They differ with regard to the applicative objective. Some works have a mostly theoretical focus, long-term works, while others have clearly applicative ends. Discourse being a particularly rich domain, there is no single approach that considers all of its aspects, to my knowledge. Therefore, the works are limited to certain aspects and ignore others, like processing anaphora, recognizing topics or analyzing discourse structure. Given this diversity, this text will present a few representative domains without claiming to be exhaustive, as that would go beyond the objectives of this book.

Finally, it should be noted that the automatic processing of discourse is particularly useful in domains like automatic translation, automatic text summarization, information extraction, etc. (see [WEB 12] for a summary of its applications).

3.2.1. *Linear segmentation of discourse*

Linear segmentation is a simple form of discourse analysis. It consists of dividing the text into parts that are deemed independent, each of which plays a precise role in the text (see Figure 3.6).

Figure 3.6 shows that the segmentation assumes the existence of clear boundaries between the parts of the text concerned. Very generally, segmentation concerns relatively large parts of the text and does not attempt to process the finer points that can only be grasped by considering the dependency relations between parts of the text.

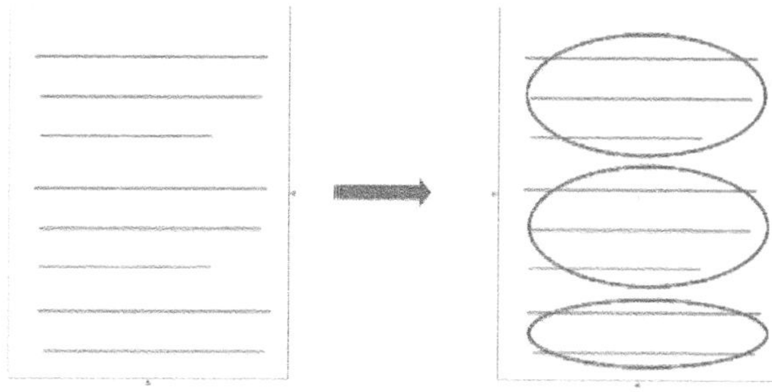

Figure 3.6. *Linear segmentation of a text*

The following sections will show in more detail how the segments can be determined based on a variety of dimensions including topics, emotions, opinions, etc.

A discourse is normally composed of a series of topics that revolve around various frameworks. Identifying these topics and their articulation within the discourse makes it possible to classify the text according to its content and its form. It is therefore normal that information retrieval is the privileged application domain of this issue (see, for example, [HEA 97, CAP 06]). In the context of spoken dialogue systems, topic recognition has been used to guide the syntactic analysis module and reduce the search space, consequently accelerating the analysis and improving the system's performance [GAV 00a]. Topic recognition has also been used to segment classes or meetings [GAL 03, MAL 06].

A coherent discourse is characterized by a logical sequencing of the topics addressed therein. Modeling this sequencing can be used to make predictions about topics that will be addressed by considering a given history or simply to identify the topics in a sequence. It should be noted that there is no standard for topics and some researchers include many different practices under the label of topic recognition. For example, some include functional segmentation based on the roles filled by the identified units or consider the intentional structure of the discourse [MOO 93, LOC 98] or the speech acts [COL 97]. Others, inspired by works on scripts [SCH 77] or Story Grammar [KIN 78, MAN 84], focus on events in the discourse and their relations [CHA 08, FIN 09].

The general principle of topic segmentation consists of dividing the discourse and using lexical indicators to predict or recognize topics. Naturally, different forms of statistical models have been adopted to model the topic triggers. Simple models based on ngrams [CAV 94], HMM [BOU 02] and Bayesian networks [SAH 98, EIS 08] have been used for similar tasks. Other techniques like Latent Semantic Analysis and neural networks have also been used [FOL 96, WIE 95]. Given the considerable similarity between topic recognition and information retrieval, the technical details will be presented in section 4.3 dedicated to information retrieval in the sphere of applications.

3.2.2. *Rhetorical structure theory and automatic discourse analysis*

Adopting a theoretical framework seems beneficial for automatic discourse analysis. Among the existing frameworks, the Rhetorical Structure Theory (RST) has been adopted in many works. Although it was initially developed with the goal of improving automatic generation systems, this theory has acquired an independent theoretical status. Intended for describing the structure of texts rather than explaining the processes that underlie their creation and interpretation, RST considers the text as a set of blocks interconnected by hierarchical relations, like coherence or discourse. From this perspective, every block fills a particular role in relation to the other blocks. An example of a text analyzed with RST is shown in Figure 3.7.

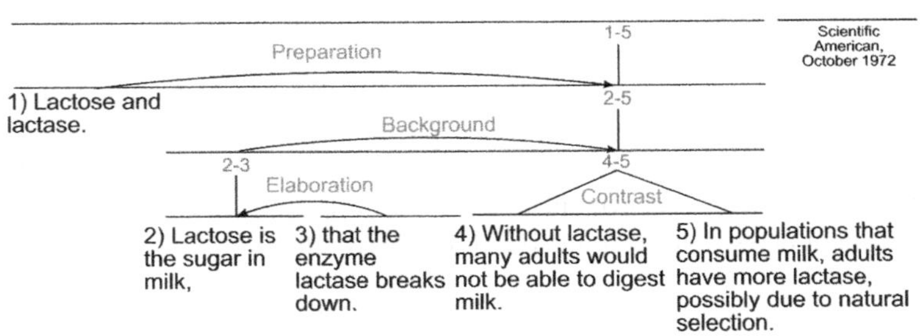

Figure 3.7. *Example of a text analyzed according to RST [MAN 12]. For a color version of this figure, see www.iste.co.uk/kurdi/language2.zip*

RST distinguishes between two types of relations: nucleus-satellite mononuclear relations and multinuclear relations. Mononuclear relations are marked by the dependency of an element called a satellite on a nucleus, whereas multinuclear relations are distinguished by the rhetorical equivalence of the elements involved. In the analysis diagrams, the root represents the set of units 1-5 of the text and the arrows, which each represent a mononuclear relation, point toward the dominant unit or block (e.g. presentation, elaboration). Multinuclear relations, in turn, are marked by two non-oriented arcs with the same root. Note that in the initial version of RST presented in [MAN 88], parallel relations can establish a relation between a discursive unit and a complex unit. More recent versions of RST explicitly require that a unique relation be present between each pair of units [STE 04, CAR 03]. Table 3.3 provides a list of mononuclear relations.

Name of the relation	Nucleus	Satellite
Anti-condition	Absence of the conditioning situation causes the occurrence of the resulting situation	The conditioning situation
Antithesis	Ideas approved by the author	Ideas rejected by the author
Background	Text whose comprehension is facilitated	Text serving to facilitate comprehension
Goal	A target situation	The intention underlying a situation
Intentional cause	A situation	Another situation that intentionally caused the first
Unintentional cause	A situation	Another situation that unintentionally caused the first
Circumstance	A text expressing the events or ideas located in the interpretive framework	A temporal or situational interpretive framework
Concession	A situation defended by the author	A situation that is apparently incompatible, but is still defended by the author
Condition	An action or situation whose occurrence results from the occurrence of the conditioning situation	The conditioning situation
Demonstration	An affirmation	Information intended to increase the reader's belief in the affirmation
Elaboration	Source information	Supplementary information
Evaluation	A situation	An evaluation of the situation
Facilitation	An action	Information intended to help the reader accomplish this action
Interpretation	A situation	An interpretation of the situation
Justification	A text	Information justifying the production of the text by the author
Motivation	An action	Information intended to encourage the reader to accomplish an action
Reformulation	A situation	A reformulation of the situation
Intentional result	A situation	Another situation, intentionally caused by this one
Unintentional result	A situation	Another situation accidentally provoked by the first
Summary	A text	A brief summary of this text
Solution	A situation or a process that satisfies a need	A need, problem or question

Table 3.3. *Inventory of mononuclear relations established in [MAN 88]*

For more clarity, examine the passages in [3.35] for examples of mononuclear relations.

[S Although he adores chemistry], [N he registered in the applied mathematics program].	concession	
[N The inspector was succinct;] [S he mentioned causes related to the confidentiality of the inspection]	elaboration	[3.35]
[S Brass is an alloy that is mainly composed of copper and zinc.] [N Add five grams of brass.]	background	

Multinuclear relations are less numerous and less semantically restrictive, except in the case of contrast, which requires an opposition between the items involved (see Table 3.4).

Name of the relation	A segment	Other segment
Contrast	a possibility	the opposite possibility
Junction	(non-constraint)	(non-constraint)
List	an item	next item
Sequence	an item	next item

Table 3.4. *Inventory of multinuclear relations established in [MAN 88]*

For example, in an enumeration, all sorts of objects that do not have inherent links can be listed (see examples [3.36]):

[Mary would like to spend the summer in Florida] whereas [John would like to go to Spain.]	Contrast	
[John is a chef in a French restaurant in New York]; [Michelle is a gymnastics teacher in Atlanta].	List (no comparison or contrast)	[3.36]

Aside from the relations between nucleus and satellite, [MAN 88] formulated constraints on the beliefs and objectives of the two actors involved, the writer (W) and their effects on the reader (R). For example, the constraints on the relation evidence are defined in Table 3.5 where N corresponds to Nucleus and *S* to Satellite.

Constraint on N	R may not believe N to a satisfactory degree for W
Constraints on S	R believes S or he will find it credible
Constraints on S+N	The reading of S by R increases his belief in N
Effect	The belief of N has increased

Table 3.5. *The constraints on the relation*

Like for syntactic analysis, discursive analysis is intended to create a complete representation of the object of analysis, which in this case is the discourse. To obtain such a representation, three actions similar to the syntactic analysis process must be conducted: segmentation, attachment, and disambiguation.

In the framework of discourse analysis systems, the text is first segmented in elementary discourse units (EDU). To avoid circularity (the choice of units based on analysis and the analysis based on units), this operation is carried out before the analysis. Note that the community does not unanimously agree about the form of EDU (see [POL 01, BAL 07] for a discussion). Behind this disagreement is a need for units that cover the entirety of the discourse. In other words, after having segmented the discourse, there should not be any segments that are not units. The problematic issues include the admission or not of discontinuities within EDU. These discontinuities are caused by dislocations or inserted sentences. To avoid the issue of segmentation, some researchers assume that segmentation has already been done by a module external to the discourse parser.

In [POL 04b], the segmentation occurs after a syntactic analysis step with lexical functional grammar (LFG). The identification of EDU is done first by segmenting the sentences into simpler units based on lexical, syntactic and semantic criteria. These units are then combined in a small tree structure equipped with a functional role that corresponds to the EDU. These units include elements such as greetings, connectives and discursive markers.

To construct a segmentation module, a syntactic analysis model is typically required first to provide the base on which the discourse segmentation will occur. The segmentation module is created by following these steps:

a) Collect or find an appropriate corpus. In this case, this is a collection of texts.

b) Annotate the corpus according to a discursive framework like RST.

c) Manually annotate the linguistic elements that can be used as boundary markers between sentences. Lexical-syntactic factors are mainly used. Another factor is prosody, which plays an important role in defining sentence boundaries [HIR 96, LEV 04] in spoken discourse.

d) Implement the linguistic rules identified in the previous sentence.

e) Evaluate the segmentation module.

After the segmentation comes the attachment phase, where the parser must decide where it will connect the current segment in the tree under construction. Here as well, there are several strategies. For example, in the parser in [POL 04b] the attachment occurs on the basis of a set of rules related to a variety of information sources:

– Syntactic information: a multitude of syntactic information is considered. For example, if the lexeme to be attached plays the role of subject or object, then the relation is one of subordination, as in the antecedent of relative propositions that can be a subject, object, or adverbial. In the case of a parallelism, it is a relation of coordination, because the coordinated elements have the same weight.

– Lexical information: the repetition of the same lexeme or the occurrences of words linked by a relation of antonymy, synonymy or hyperonymy, as well as discursive connectives, adverbs and temporal indicators (aspect, tense and mode of the verb) all contribute to the attachment. For example, connectives like *in that case*, *therefore* and *thus* make it possible to predict a change in the discursive content and consequently the boundary of a discourse segment (for a detailed discussion, see [POL 04a]).

– Structural information: this concerns the point of attachment of the current unit in the tree under construction.

– Presence of incomplete construction constituents: this includes questions, opening greetings, internal units like sections and sub-sections, etc.

The weight attributed to these different knowledge sources varies depending on needs. For example, lexical knowledge has more importance for the point of attachment while semantic and syntactic knowledge is more relevant for choosing the relation type.

To move from the syntactic structure provided by the syntactic parser to a discursive structure, [POL 04b] uses a set of symbolic rules. Some examples are shown in Figure 3.8.

1) BDU-1, BDU-2[2]:

   BDU-1/phi = BDU-2/ADJUNCT/link;

   → Right -Subordination-.

2) BDU-1, BDU-2:

   BDU-1/*/ADJUNCT/link = BDU-2/phi;

   → subordination.

3) BDU-1, BDU-2:

   BDU-1/*/{XCOMP|COMP|COMP-EX}/link

   = BDU-2/phi;

   → Context.

Figure 3.8. *Example of rules used to construct the discursive structure starting from the syntactic structure [POL 04b]*

The first rule describes the case of prepositional and adverbial groups that are often temporal modifiers that precede the main clause that they modify. These groups can either elaborate on the content of the main clause or modify the context of its interpretation. Lexical information is used to distinguish between several types of modifiers. The second rule describes the case of the subordination of an adjunct clause. They show that syntactic rules allow for a recursive search in the functional structures. Finally, the third rule expresses the disjunctive constraints that construct a relation at the level of discourse.

A simpler form of discourse analysis called chunking or partial analysis was also proposed in simulation to the partial syntactic analysis proposed by Abney (see for example [MID 03, SPO 05]). The main difference from the complete analysis just presented resides in the fact that the chunking process seeks to recognize segments or islands in the discourse, without attempting to analyze everything. Thus, a chunker will focus on identifying morphemes, terms and constructions that will help to indicate the discursive relations in a given context. For each of the identified items, the chunker attempts to find its arguments and to determine the type of semantic relation that connects them.

2 BDU stands for Basic Discourse Unit, and the asterisk is a wildcard that accepts any element.

In terms of applications, the structural analysis of discourse has proven to be particularly promising in several applicative domains at the forefront of which is automatic text summarization [ONO 94]. More recently [MAR 00, THI 04] proposed an approach based on a varied weight attribution process for different components of a discursive tree.

3.2.3. *Discourse interpretation: DRT*[3]

The previous section discussed the construction of a tree structure for discourse. This level of analysis can easily be considered the syntax of the discourse. Naturally, every syntactic structure needs a semantic framework for its interpretation and that is where Discourse Representation Theory (DRT) comes in.

In modern semantics, a new movement emerged. The focus of this movement is discourse, which it considers to be the unit that must have a truth value, rather than the sentence. This movement was initiated by DRT, developed by Hans Kamp [KAM 81], with the goal of accounting for discursive phenomena like anaphora and tense, whose representation goes beyond the framework of predicate logic. A similar work was realized by Irene Heim called File Change Semantics (FCS) [HEI 82]. The works of Gilles Fauconnier on mental spaces are also considered in the same tradition as DRT [FAU 84]. Consider sentence [3.37] and the corresponding logical representation. A human being is naturally capable of guessing that the anaphor *he* refers to Max. For a computer program, a post-processing algorithm is required. Such an algorithm is intended to replace the free variable *X* with the appropriate antecedent and provide the correct representation:

Max is a cat. He likes Felina.

cat(max) ∧ likes(X, Felina) → [X is a free variable] [3.37]

cat(max) ∧ likes(max, felina)

To obtain representations of this type, DRT uses representations called discourse representation structures (DRS), which go beyond Montague semantics, which is limited to the sentence level. The discourse representations are incrementally enriched, similar to what a human being would do in receiving (listening or reading) a sequence of sentences. Illustrated representations in the forms of boxes are used for these structures following the example of the tradition that is popular in the domain of the psychology of language. Each box is composed of two parts: an upper part that contains the referents and a lower part that contains the conditions. The

3 Discourse representation theory (DRT).

referents are the set of entities that were introduced in the context, while the conditions are the predicates that connect these entities.

To construct a representation of [3.37], it is first necessary to introduce the two basic elements in the discourse, which are: x, y, max, likes(y, Felina) in two different boxes, which correspond to the two sentences in the discourse. Then, the final representation is obtained with a juncture of the referents and the conditions of the two sentences (see Figure 3.9). Note that the equality of the symbols x and y is the way to express the fact that the two symbols refer to the same entity, thus identifying the referent of the anaphor. The DRT creates a temporary representation of the discursive units that it updates each time that new entities are introduced into the discourse. In other words, each sentence is interpreted in a given context. The result of the interpretation of this sentence contributes to updating the context and interpreting the other sentences and so on.

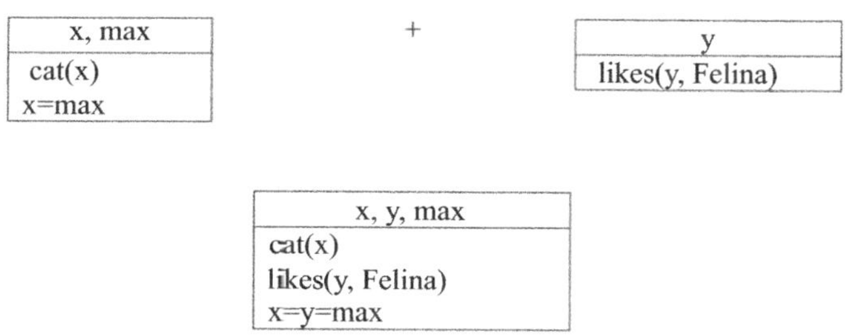

Figure 3.9. *DRT representation of [3.35]*

Consider utterance [3.38] with quantifiers as well as its logical representation:

Every farmer who owns a donkey feeds it. [3.38]

Contrary to Russel's understanding, the indefinite article of the group *a donkey* must be replaced by a universal quantifier. Two boxes are used by the DRT to represent the two propositions in this sentence (see Figure 3.10).

In the context of natural language processing, it is possible to construct discursive representations according to DRT using algorithms. To show this, we will use the top-down approach presented in [BLA 05]. Consider example [3.39] to illustrate this algorithm:

A man yells. He dies. [3.39]

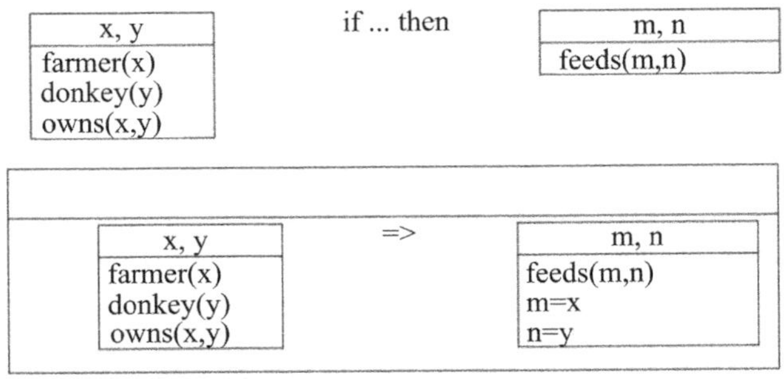

Figure 3.10. *DRT representation of [3.35]*

The first step of the algorithm is to construct the skeleton of the representation from the root of the tree corresponding to the first sentence of the micro-discourse (see Figure 3.11).

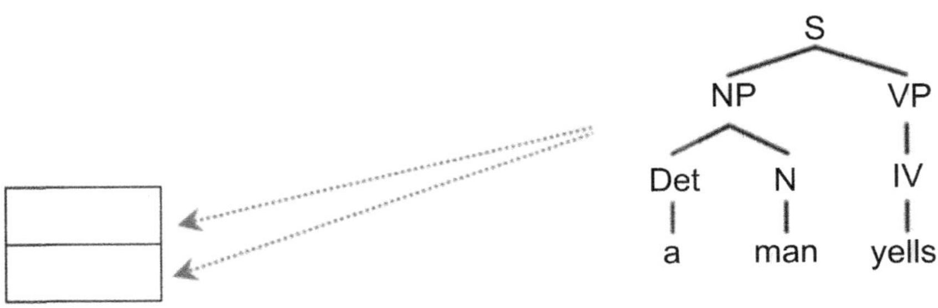

Figure 3.11. *First step of the algorithm*

To begin to populate the representation, proceed with a top-down exploration from left to right. This means visiting the symbols: S → NP → DET. Because there is an indefinite nominal group, a new object x is added in the upper part of the box (see Figure 3.12).

Figure 3.12. *Second step of the algorithm*

Continuing to explore the nominal group, we discover that it consists of a man. This fact is expressed in the form of a condition in the lower part of the box: man(x) (see Figure 3.13).

Figure 3.13. *Third step of the algorithm*

After having finished exploring the entirety of the nominal group, move on to the verbal group that includes an intransitive verb. This leads us to add a new constraint on x: yell(x) (see Figure 3.14).

Figure 3.14. *Fourth step of the algorithm*

The representation of the first sentence being complete, move on to the second sentence. By following the same approach described previously, start with the root of the tree (Figure 3.15).

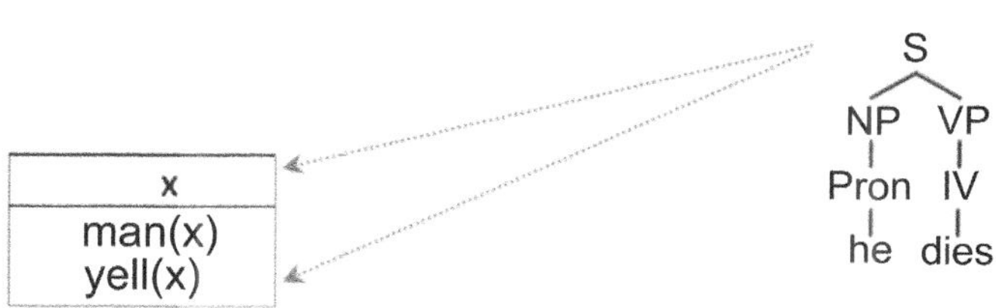

Figure 3.15. *Fifth step of the algorithm*

To reach a unique representation of the discourse, first add the representation of the indefinite nominal group that is a new variable, y. The anaphora resolution occurs by admitting the equality of the two referents x and y. This equality is decided on the basis of the person (third), the number (singular) and the gender (masculine) (Figure 3.16).

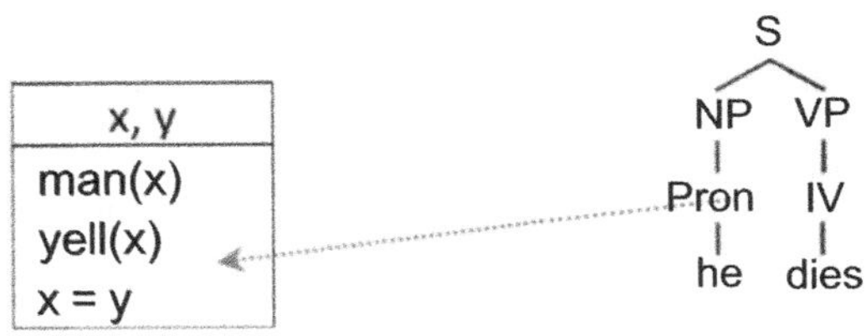

Figure 3.16. *Sixth step of the algorithm*

Finally, the representation of the verbal group, the intransitive verb *dies*, occurs with the condition dies(y). The semantics of the verbs *yell* and *die* create a logical sequence which, although a bit simplified in this example, confirms the equality decision between the two referents (Figure 3.17).

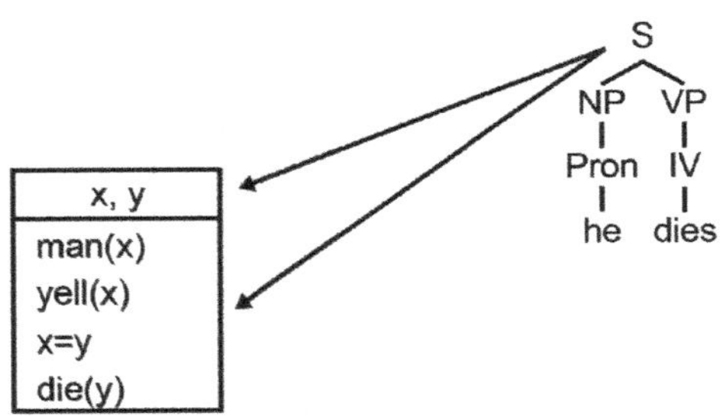

Figure 3.17. *Seventh step of the algorithm*

In summary, DRT makes it possible to reach several objectives. Thanks to the mechanism that updates temporary representations, it allows for the elegant processing of indefinite determinants and cases of presupposition. The identification of the unique denotation of a determinant makes it possible to process new entities introduced in the discourse as well as quantified variables. It is also an elegant way to account for anaphors, especially those that refer to quantified variables.

Similarly, according to the approach initiated in works on dynamic semantics (DS) [GRO 91], fragments of text or discourse are considered to be instructions to update the existing context. Contrary to DRT, this does not consist of updating the representation of the semantic content, but rather of updating the content itself [DEK 12].

It should be noted that DRT has been the object of large-scale applications, notably in the context of the Verbmobil project for the translation of spontaneous spoken dialogues [WAH 95, BOS 96].

3.2.4. *Processing anaphora*

Since the beginning of applications in the domain of Natural Language Processing, researchers have recognized the interest in processing phenomena like anaphora. Thus, applicative systems like STUDENT, an Intelligent tutoring system [BOB 64] and SHRDLU, a natural language command interface [WIN 72] have integrated anaphora resolution functionalities. Similarly, algorithms specifically dedicated for this task were created, such as those of [HOB 76, HOB 78].

The question of processing anaphora consists of resolving the challenging problem of finding their referents. Indeed, without identifying the referent, it is not possible to find an adequate semantic representation for certain sentences. Two criteria can be used to distinguish the different algorithms and approaches for processing anaphora: information resources and processing method. Most anaphora resolution algorithms count on linguistic information sources like agreement in gender and number, syntactic dependency, etc. There are two competing currents with regard to the processing method. On the one hand, there are knowledge-based approaches where system designers must provide a detailed and exhaustive description of phenomena to be covered. On the other hand, there are learning-based approaches which are generally based on real data that is manually annotated (see [DEO 04, MIT 98, WIS 16] for a general overview of these algorithms).

3.2.4.1. *The naive syntactic approach of Hobbs*

Syntactic approaches are used to filter a pronoun's antecedent candidates. As noted, syntactic factors are important but not sufficient to determine an antecedent. The basic idea of Hobbs' algorithm [HOB 76, HOB 78] is to move through the syntactic parsing tree of a sentence searching for a nominal group that satisfies predefined constraints like number and gender. More specifically, the search is carried out according to the following steps [HOB 78]:

1) Start with the node of the Nominal Group *NP* that immediately dominates the pronoun.

2) Move higher in the tree until the first node *NP* is encountered or until the node of the sentence *S*. Call this node *X* and the route taken to find it *R*.

3) Go through all of the branches above *X* and to the left of the route *R*, following a breadth-first left-to-right manner. Propose as an antecedent any node *NP* encountered that has a node *NP* or *S* between it and *X*.

4) If *X* is the highest node *S* in the parsing tree, then go through the parsing trees of preceding sentences in chronological order, with the most recent sentence first, always from left to right and breadth first. When a node *NP* is encountered, propose it as antecedent. If *X* is not the highest node *S* in the sentence, then move on to phase 5.

5) Go from *X* until the first node *NP* or *S* is encountered. Call this new node *X* and the route to it *R*.

6) If *X* is an *NP* and if *R* does not cross the node *N'* directly dominated by *X* then propose *X* as the antecedent.

7) Go through the branches above *X* to the left of the route *R* following a breadth-first left-to-right manner. Propose any *NP* encountered as an antecedent.

8) If *X* is a node *S*, then go through all the branches of *X* to the right of *R* following a breadth-first, right-to-left manner, but without going above any *NP* or *S* encountered. Propose any node *NP* encountered as an antecedent.

9) Go to step 4.

Although this algorithm is rather basic, a comparison with systems such as those described in [BAL 97, MIT 97, TET 99] show that it is not entirely obsolete. In addition, as noted by Hobbs, its low cost computationally and its simplicity of implementation make it appealing even today, especially in the context of applications that require processing in real time. This explains the adaptation of this algorithm to other languages (see for example [DUT 08]).

3.2.4.2. *The discursive approach: centering theory (CT)*

The theory of Grosz and Sidner distinguishes between three main components of discourse [GRO 86]: the linguistic structure, the intentional structure, and an attentional state. Regarding the linguistic aspects, discourse is seen as a set of segments and relations relating the pairs of segments. The intentional structure in turn includes the communicative intentions and relations between them. The attentional states concern the focus of the attention of the participants in the discourse at a given moment. Two types of attentional states are distinguished here: local and global. The first concerns relations between utterances in a discourse segment that contribute to the creation of a local coherence (intrasegment). The second concerns the relations between the segments within the discourse (intersegment). Changes in the attentional state depend on both the intentional structure and the linguistic structure.

The foundational ideas of CT were discussed in the works of [GRO 77, SID 79, ARA 81] and concern the semantic links between an utterance and the preceding and subsequent utterances in the context of a discourse segment for local relations. Among others, these ideas pertained to the levels of focus in a discourse, the possible centers of focus in a discourse and using them to choose the referent of an anaphora, and finally the relations between the change of focus and the complexity of inferences necessary for the integration of the current utterance with the rest of the discourse. The working hypothesis is that at any moment, there is a single unit (among the units evoked in the discourse) that takes center stage. This is the centered unit. To clarify this principle, consider passages [3.40] and [3.41]:

a) John helped the truck driver Michael to find the street where the factory was.

b) He looked at a map of the area while Michael drove. [3.40]

c) He finally managed to find the location.

a) John helped the truck driver Michael to find the street where the factory was.

b) He looked at a map of the area while Michael drove. [3.41]

c) Suddenly, he braked when he saw a small cat crossing the street.

In passage [3.40], the pronoun *he* in utterance c refers to John because it is logical to infer that the person looking at the map will manage to find the location. Similarly, the action of driving the truck normally excludes looking at a map. The anaphor *he* in passage [3.41] logically corresponds to Michael because the realization of the action of the verb to brake presupposes the action of driving.

According to Hobbs' syntactic approach, passages [3.40] and [3.41] are perfectly equivalent in terms of coherence. On the other hand, according to CT, passage [3.40] is more coherent than passage [3.41]. Indeed, as Michael is the center of the utterances a and b, the change of the center in passage [3.41] makes inferring the referent of the anaphora more challenging and consequently decreases the coherence of the discourse.

More formally, according to the model proposed by CT, discourse is composed of a set of segments, each of which represents a part of the discourse model. The centers are semantically prominent entities that are part of the discursive model of each utterance E_i in a discourse segment (DS) that consists of a series of utterances: E_1, E_2, ... E_m. Each utterance is associated with two centers: forward-looking centers and backward-looking centers.

Forward-looking centers C_v (E_i, DS) represent the set of entities evoked by the utterance E_i in a DS. These entities can be repeated by subsequent entities in the DS and are ordered according to their prominence. In other words, the more highly ranked an entity is, the more chance it has of being repeated. The order of elements is determined by a number of factors such as the grammatical role of the entity, the order of the entity and the informational *novelty* that it conveys according to Prince's criteria [PRI 81]. For example, in passage [3.40]: $C_v(E_a)$={John, Michael, street, factory}. The highest ranked entity is called the Preferred Center C_p. Thus, the preferred center is John: $C_p(E_a)$ =John. The C_p serves to predict the backward-looking center C_r of the next utterance.

The backward-looking center $C_r(E_i$,DS) represents the most highly ranked discourse segment for U_i, among those evoked by the previous utterances in the DS. It has the role of connecting an utterance to a previous utterance, which is in turn connected to a previous one and so on. Based on the results of several psycholinguistic studies, CT stipulates that a given utterance U_i cannot have several C_r at one time [HUD 88, GOR 93]. For example, in passage [3.40], the backward-looking center of the utterance a is undefined: Cr(Ea)=[?]4, because this utterance is the first in the DS. However, the backward-looking center for utterance b is John: Cr(Eb)= John (the highest ranked entity in utterance a cp(Ea)=John or Cv(Ea)).

Aside from the centers just presented, CT includes the following two rules concerning the center realization and movement: the pronominal rule and the transition specification rule.

The pronominal rule stipulates that pronominalization is a possible way to mark the prominence of an element and that the Cr are often removed or pronominalized. Consequently, if there are a multitude of pronouns in an utterance and if they correspond to several entities in the previous utterance, then the Cr must be one of these pronouns. Naturally, if there is only one pronoun in an utterance, this pronoun must correspond to the Cr. More formally, this rule asserts that if an entity among the ones listed in Cv(Ei-1, SD) is realized in the form of a pronoun in Ei, then Cr(Ei, SD) must also be a pronoun.

The transition specification rule concerns relations between adjacent utterances. It stipulates that the different types of transition require variable quantities of inferences, thus causing variations in levels of coherence. The transitions that require fewer inferences correspond to a more coherent discourse, where there is minimal change to the center. Four types of transitions are described in [BRE 87].

4 For utterances at the start of a discourse, cases where the C_r is undefined are marked *undefined* or [?].

The order of preference of the transitions is: continuation > retention > smooth shift > rough shift (see Table 3.6 for a summary).

	$C_r(E_{n+1}) = C_r(E_n)$ or $C_r(E_n) = [?]$	$C_r(E_{n+1})$!= $C_r(E_n)$
$C_r(E_{n+1}) = C_p(E_{n+1})$	Continuation	Smooth shift
$C_r(E_{n+1})$!= $C_p(E_{n+1})$	Retention	Rough shift

Table 3.6. *The four types of transition possible in a discourse segment*

Continuation occurs when a locutor indicates that he intends to discuss a particular topic. This implies that the center of an utterance has a good chance of spreading not only to the next utterance but also to the rest of the utterances in the discourse segment. In other words, in the case of a continuation: $C_r(E_{n+1})=C_r(E_n)$ or $C_r(E_n)=[?]$ for E_n at the start of a discourse and $C_r(E_{n+1}) = C_p(E_{n+1})$. It is also very likely that $C_r(E_{n+2})=C_r(E_n)$.

In the case of retention, the locutor repeats the center of an utterance from one utterance to another but seems to want to change the center in the ensuing utterances. This translates into the fact that the current backward-looking center and the next utterances are the same, but the center is not the most highly ranked entity among the ones listed. More formally, this gives: $C_r(E_{n+1}) = C_r(E_n)$ or $C_r(E_n) = [?]$ for E_n at the start of the discourse and $C_r(E_n)$!= $C_p(E_{n+1})$. This means that it is very likely that $C_r(E_{n+2})$!= $C_r(E_n)$.

With a smooth shift, the locutor turns his attention toward a new entity about which he intends to continue speaking. In this case, $C_r(E_{n+1})$!= $C_r(E_n)$ and $C_r(E_{n+1}) = C_p(E_{n+1})$.

In the case of a rough shift, like a smooth shift, the locutor turns his attention toward a new entity. However, it is very likely that this new entity will not be the focus of the utterances that follow. Therefore, $C_r(E_{n+1})$!= $C_r(E_n)$ and $C_r(E_{n+1})$!= $C_p(E_{n+1})$.

To illustrate these transitions, examine passage [3.42]:

E1 The genie has lived in the magic lamp for thousands of years.

E2 He currently serves Aladdin.

E3 He promised him he would find the lost treasure. [3.42]

E4 He does not believe him.

An analysis of passage [3.42] utterance by utterance according to CT gives the steps presented in Figure 3.18.

1) Generation of possible combinations of centers C_r-C_p for each utterance.
2) Filtering candidates using syntactic constraints, selection constraints, rules, and CT constraints.
3) Sequencing following the transition priorities.

Figure 3.18. *Algorithm for processing pronominal references [BRE 87]*

The algorithm in [BRE 87] proposes processing pronominal references according to the principles just described (see Figure 3.19).

E_1 The genie has lived in the magic lamp for thousands of years.
$C_r(E_1)$ = [?] *utterance at the start of the discourse*
$C_p(E_1)$ = the genie
$C_r(E_2) = C_p(E_2) \wedge C_r(E_1)$ = [?] → Transition = continuation
E_2 He currently serves Aladdin.
$C_r(E_2)$ = the genie
$C_p(E_2)$ = the genie
He= the genie
$C_r(E_3) = C_p(E_3) \wedge C_r(E_3) = C_r(E_2)$ → Transition = continuation
E_3 He promised him he would find the lost treasure.
$C_r(E_3)$ = the genie
$C_p(E_3)$ = the genie
He = the genie; him = Aladdin *we assume that the genie made the promise to Aladdin*
$C_r(E_4) \mathrel{!=} C_p(E_4) \wedge C_r(E_4) = C_r(E_3)$ → Transition = Retention
E_4 He does not believe him.
$C_r(E_4)$ = the genie
$C_p(E_4)$ = Aladdin *we assume that Aladdin does not believe the genie*

Figure 3.19. *Transition of centers in example [3.39]*

A generalization of CT called Veins Theory (VT) was proposed by [CRI 98]. The basic idea of this theory consists of identifying the veins in discursive trees. The role of these veins is to define the domains of the referential accessibility of each unit in the discourse. It should be noted that CT has been applied in the context of automatic scoring systems for written texts based on the distinction between the types of transitions, like in the e-rater system [MIL 00].

From a psycholinguistic point of view, several works have confirmed the cognitive relevance of CT principles. The studies in [CLA 79] found that the reading time for passages where the referent is found in the preceding utterance are shorter than in passages where the referent is found two or three utterances back. Similarly, [CRA 90] found that subjects managed to identify the referent more quickly when it was a grammatical subject than when it was an object.

3.2.4.3. *Learning-based approaches*

Learning-based approaches use a different philosophy to solve the problem. Rather than manually coding linguistic rules that describe syntactic, semantic and discursive patterns, they annotate representative corpora. The linguistic knowledge is therefore reduced to a minimum and the parameters of the final model are automatically induced. Typically, anaphora and/or their possible referents are annotated with features such as gender, number, grammatical role (like subject or object) or linguistic form (like proper noun or definite or indefinite nominal group). In order to coordinate efforts in the domain of the annotation of corpora intended to train algorithms to process anaphora and coreferences, several standards were created, including MATE[5] [POE 04, POE 99].

Bayesian classifiers and recurring neuronal networks have been used with the same objective [HUA 13, FRA 08, WIS 16], but the basic classifiers of Decision Trees (DTs) were the most commonly used framework for these tasks [MCC 95, KWA 08, SOO 01, PAU 99] until the recent return to neuronal network-based methods. Simplicity and performance are the reasons cited to justify the use of Decision Trees for classification. The fact that these trees generate rules makes them more appealing for linguistic applications. Historically, several versions of DT algorithms were proposed, like ID3, C4.5 and C5.1.3 [QUI 79, QUI 93, KOH 99]. In the DTs employed for anaphora resolution, the leaves represent the classes and the branches correspond to the conjunction of features that lead to the leaves and therefore the classes. [MCC 95] described a comparison of DTs with a handwritten rule-based system. The work was realized on the English Joint-Venture (EJV)

5 Multilevel Annotation Tools Engineering.

corpus. This is a corpus that is rich in coreferences from projects realized by several entities: government, private companies, individuals, etc. The list of features used in the corpus annotation includes (see Figure 3.20 for an overview):

– NAME i: does the reference *i* contain a name? The possible values are *Yes* and *No*.

– JV-CHILD i: does the reference *i* refer to a *Joint Venture* child? For example, a company or a project resulting from a partnership between two entities. The possible values are *Yes*, *No* and *Unknown*.

– ALIAS: does one reference contain an alias for another? In other words, is the name of a reference a substring of the name of another? The possible values are *Yes* and *No*.

– BOTH-JV-CHILD: do both references refer to a JV child? Three values are possible: *Yes*, *No*, *Unknown*.

– COMMON-NP: do the two references share a common Noun Phrase (NP)? These are generally references to a complex nominal group, as in the case of relative propositions and appositions. This feature compares the simple constituent NPs to each reference. The possible values are *Yes* and *No*.

– SAME-Sentence: do the references appear in the same sentence? *Yes* and *No* are the possible values for this feature.

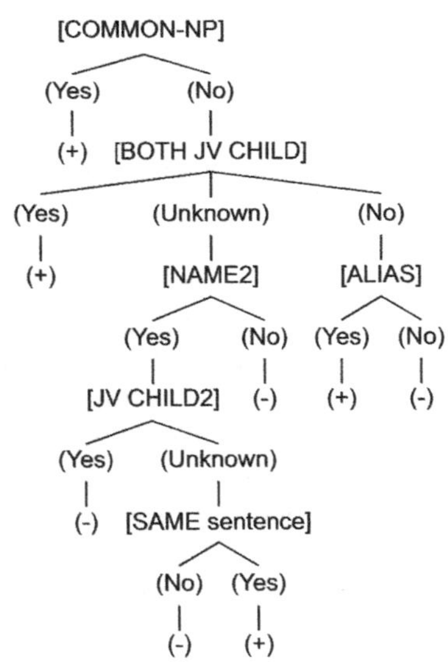

Figure 3.20. *Decision tree C4.5 for coreference resolution [MCC 95]*

The features in the example can have three possible values at maximum: *Yes*, *No*, *Unknown*, while only two classes are possible in this tree: coreference (+) and absence of coreference (-). Note that the interest of decision trees obtained by automatic processing, compared to rules written by linguists, resides in the fact that these rules are automatically induced according to criteria related to the gain of information and are therefore more optimal compared to the content of a corpus used for learning. Moreover, these trees have a much lower production cost. To understand these differences, examine the tree construction algorithm presented in Figure 3.21.

```
If all cases (in the corpus) belong to the same class C
        Then the result is a tree with C as the root
If not,
        Choose the most informative attribute A, whose values are v1 … vn
        Divide the cases in the corpus into groups G1 ... Gn based on the values of A
Recursively construct the sub-trees T1 … Tn for G1 ... Gn respectively
The final result is a tree whose root is A and whose trees are T1 … Tn .The branches
between A and the T1 … Tn are labeled by the values v1 … vn
```

Figure 3.21. *Basic algorithm for creating decision trees [QUI 79]*

The skeleton of the decision tree to be constructed will have the form given in Figure 3.22.

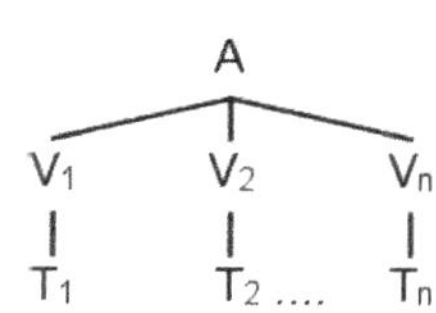

Figure 3.22. *Skeleton of a decision tree*

The choice of the most informative attribute is the most interesting aspect in the DT construction algorithm. To do this, criteria based on information theory are used, starting with the concept of entropy that indicates the degree of impurity or the disorder of the data in the database in the case formalized by the equation [3.43] [SHA 48]. In this equation, we assume that an attribute *A* takes *n* values while p(i) is the probability that *s* belongs to the category *i*. The attributes will make it possible to share the data in the corpus in subsets according to the values of *A*:

$$\text{entropy}(s) = -\sum_{i=1}^{n} p(i) \log_2 p(i) \qquad [3.43]$$

In this algorithm, what we are looking for is not exactly entropy, but the attribute that reduces this entropy. To do this, we use another measure that depends directly on the entropy, the information gain, whose formula is given in [3.44].

$$Ires(A) = -\sum_{v} p(v)\sum_{c} \frac{p(v,c)}{p(v)} \log_2\left(\frac{p(v,c)}{p(v)}\right) \quad [3.44]$$

In equation [3.42], v corresponds to the values of A, p(v) is the probability of the value v in the set S and p(v, c) is the probability that an object will be in class C and that it will have the value v. The probabilities p(v) and p(v, c) are statistically calculated according to the occurrences in the corpus.[6]

The comparison of the F-measures (see section 4.1.9.2 for an introduction) by McCarthy between the handwritten system and the computer system shows that the system that uses DTs gives better results than rule-based systems, at 86.5% and 78.9% respectively. Despite their advantages in terms of results, the major inconvenience of learning-based systems is the necessity of having good-quality corpora with appropriate annotation. Unfortunately, such corpora are not always available, especially for languages with few linguistic resources.

6 For lack of some particular information about an event, the best estimator of its probability is its relative frequency in the space of observables.

4

The Sphere of Applications

4.1. Software engineering for NLP software

In the previous chapters, the basic components of NLP systems were presented. They focused on separate linguistic levels such as speech, morphology, syntax, semantics and discourse. This chapter will focus on combining different levels of linguistic knowledge with other sources of knowledge to build applicative NLP software that is directly usable by humans. Particular attention will be paid to presenting the specificities of NLP software regarding their development cycle, architecture and evaluation.

4.1.1. *Lifecycle of an NLP software*

Developing an NLP software is a complex process that involves many actors and occurs through successive or parallel steps, some of which are optional. It all starts with the identification of a need, whether it comes from the world of research, the daily practices of the general public or a business opportunity. Then comes the analysis and definition of the system prerequisites recorded in the project specifications. The next step is the design, which consists of describing all aspects of the system, including the interface, creating diagrams, writing documentation and pseudocode. During the development phase, the system's code is written, generally in a modular way. A test of the system's code is then carried out. The modules are tested individually (modular tests) and when they have been integrated into the system (integration tests). The development phases and the tests therefore partially overlap. The final phase of the development process is maintenance. This is a phase that consists of verifying that the system always responds to prerequisites and identifying points that can be improved. In the case of NLP systems, there is often an additional phase of data collection. The steps and the approaches of such a procedure are described in the spheres of linguistic resources in the first volume of this book, and this step occurs between the steps of designing and writing code.

Several models for managing software development phases have been created, including the spiral model, the incremental model and the Extreme Programming (XP) model (see [BEL 05] for more details about these different models). The rapid prototyping approach is probably the most commonly used option for developing NLP systems. With this approach, the development of the system occurs according to several large steps, each of which involves most of the phases presented here. It starts with a basic prototype that covers a part of the domain to evaluate its compatibility with the project specifications. An update to the design and then the code produces a new version of the prototype until the final version of the system is produced. A review of some software engineering questions regarding the development of linguistic software can be found in the first part of the dissertation of [LAF 94] as well as in [LEI 03].

The boundaries between linguistic levels are not absolute. In many cases, to solve a problem at a given level, knowledge of the higher level is needed. This is especially the case with the problem of ambiguity, which is very often lexical. There are two strategies to deal with this. The first consists of making a decision based on the available knowledge and sending the most correct possible result to the higher-level modules. The advantage of this strategy is that it is the simplest to implement, but the results rendered may not be optimal. The second strategy consists of involving the higher levels in the decision-making. This occurs either by invoking the higher levels to solve the problem or simply by propagating the ambiguity as is in the higher levels.

4.1.2. *Software architecture for NLP*

In the early 1950s, programs were written in machine language. At the time, the instructions and the data were individually and explicitly placed in the computer's memory. To carry out even the slightest modification of the code, a manual verification of the entire code was often necessary. Later, the first high-level programming languages started to be created. Toward the end of the 1960s, experienced programmers intuitively realized that organizing the information in the data structures would make programming easier. In the 1970s, these insights were realized as a formal theory of software architecture.

Software architecture is the structure of the system that includes the software components (modules), the properties visible on the exterior of these components and the relations between them [BAS 98]. The interest in a description of the software architecture is manifold. It makes it possible to determine the relevance of the software design in relation to the prerequisites and to discuss alternatives, thereby reducing risks of error in the development process. A description of the software architecture also sometimes makes it possible to anchor the software in the context of works originating from other related disciplines, including the cognitive sciences in particular. Finally, a well-designed architecture makes it possible to divide the system into units or modules that are likely to be reused in other projects. Similarly, this makes it possible to distribute tasks within the project team, which can sometimes be widely distributed geographically. Data flow diagrams (DFDs) (see [KEN 13] for an introduction) are commonly used to describe the software architecture of systems the same way that some UML diagrams are, especially in the context of object-oriented applications (see [RUM 04] for an introduction to UML).

Several forms of software architecture have been developed, many of which are intended for the development of commercial applications for information systems such as the famous layered architecture. The following sections are limited to architectures that are most relevant for NLP. For more details about software architecture in general, see [GAR 93, GOR 06].

4.1.3. *Serial architectures*

Serial architectures are the simplest. They have two types of constituents: modules and connections. In the case of NLP systems, the modules can do the processing of linguistic components (morphological analysis, syntactic parsing) or accessory modules such as databases, ontologies, etc. The connectors make it possible to transmit data from one module to another. They can be unidirectional or bidirectional. According to the principles of this architecture, the modules must be independent entities without any sharing of state or data.

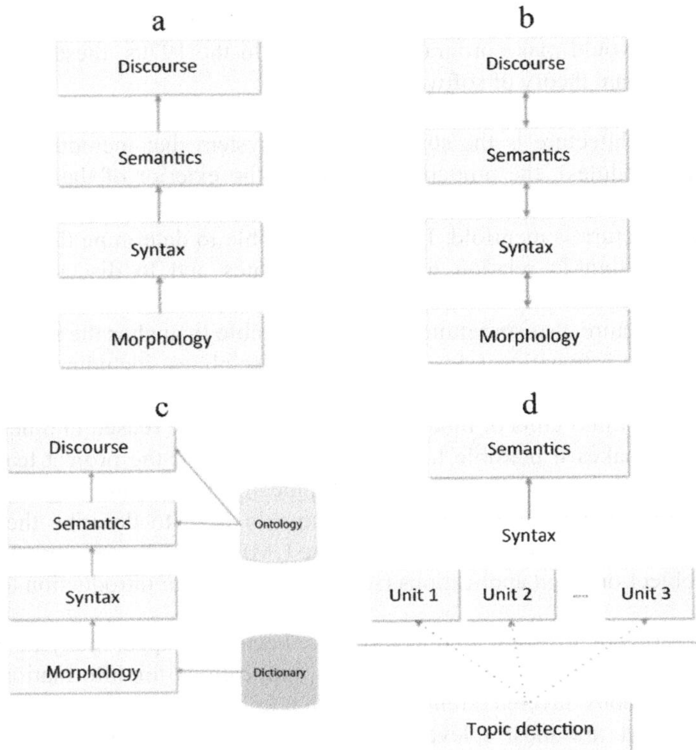

Figure 4.1. *Examples of simple serial architectures. For a color version of this figure, see www.iste.co.uk/kurdi/language2.zip*

Figure 4.1 presents four examples of simple architectures. Type (a) is a bottom-up unidirectional architecture, where the information moves from the lower-level modules to the higher levels. Decision-making occurs only at the local level, because higher-level knowledge is not invoked for the decision-making. Type (b) is a bidirectional architecture, where each module exchanges information with the neighboring modules, notably to resolve ambiguities. Type (c) is a bottom-up architecture that uses external data, in this case a dictionary, for the morphological analysis module and an ontology for the semantic and discursive analysis modules. Type (d) has a conditional activation phase. Based on a topic detection module, this architecture only activates syntactic parsing units that are relevant for a given context. This has the advantage of only mobilizing the grammatical and lexical resources that are strictly necessary for the processing in progress, which contributes to considerably improving the efficiency of the system without affecting the quality of the processing (for more details about this kind of architecture, see ([GAV 00b, KUR 03]). In all of these cases, it is assumed that the input into the system is a sentence or a text and the output is high-level linguistic analysis. To facilitate the integration of modules, the adoption or specification of standards, as much for the format as for the content, seems to be a good choice. XML and JSON are particularly useful data formats for this kind of task.

Aside from their simplicity, serial architectures have the advantage of facilitating reusability and a better understanding of the global behavior of a system. The implementation of modules can occur in parallel, and each team can focus on developing a module independently of the other teams' work. This kind of architecture facilitates the comparison of several modules. For example, we can use two or more morphological analysis modules and compare the effects of the change on the global performance of the system.

Serial architectures are not very well suited to process applicative domains where the interaction between the modules is very complex. However, many NLP applications require such an interaction. Because of the reduction of interactions between modules, this architecture encourages local decisions that only consider a limited part of the information sources available. This pushed researchers to explore architectures that offer more possibilities, such as blackboard or hub architectures.

4.1.4. *Data-centered architectures*

The idea behind these architectures is to create a central space shared by all the modules. Two main types of spaces have been used: blackboards and hubs.

In blackboard-based architectures, there is a central module that plays the role of active data repository, surrounded by a number of peripheral modules. In addition to storing data to be shared, it contacts all of the modules to inform them of the existence of these data or their state change. In other words, the blackboard plays the role of a memory space shared by the system modules.

The first implementation of a blackboard was realized at the end of the 1970s in the context of a speech recognition system called Hearsay II [ERM 80, ERM 81]. The content of a blackboard can be seen as hypotheses, which are often hierarchically structured. The modules, seen as knowledge sources, only communicate through the blackboard (see Figure 4.2).

The solicitation of knowledge sources occurs through a control module. This module can be implemented as an integral part of the blackboard, in other modules, or as a completely separate module.

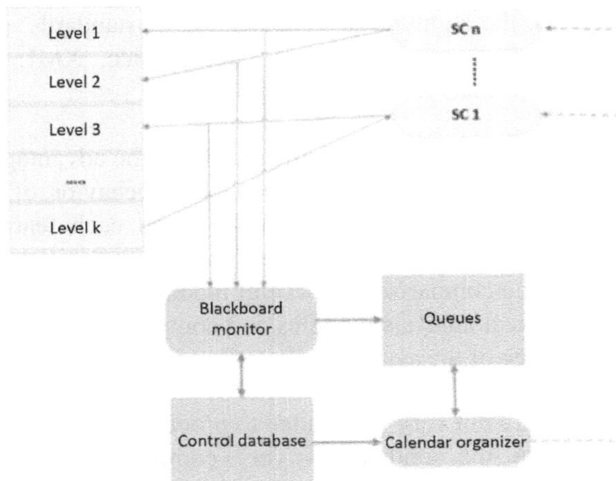

Figure 4.2. *Architecture of the Hearsay II system*

In Hearsay II, the blackboard is divided into a set of information levels that correspond to the intermediary levels of representation of the acoustic decoding process (phrase, word, syllable, etc.). Each hypothesis is located at the level of corresponding knowledge and is labeled with additional information about managing the analysis process such as the time frame within an utterance and the confidence value. The possible hypotheses at a given level constitute a search space for each knowledge source of the level under consideration. This space can be considered in a top-down or bottom-up fashion.

Figure 4.3 shows that the blackboard of the Hearsay II system is composed of six levels: acoustic parameters, from which segments are produced, the syllables that are used to make word hypotheses, which are the basis for predicting word sequences, which are phrase hypotheses. The VERIF operation attributes a weight to the coherence of the word hypotheses and contiguous word segments. The CONCAT operation creates a phrase hypothesis from a pair of contiguous word-groups. The top level is that of semantics, which generates an unambiguous interpretation of the user's query. All the hypotheses in the blackboard have a confidence value *CONF*. The *PREDICT* process predicts all of the words that can follow or precede a given group. The *STOP* process decides to stop the processing when it detects a sentence with an acceptable confidence value or when the input data has been entirely processed. A blackboard-based recognition system has been realized at the ICP in Grenoble [NAS 90]. Similarly, this kind of architecture has been adopted in domains beyond the classical applications of artificial intelligence, for example, web applications [MET 05].

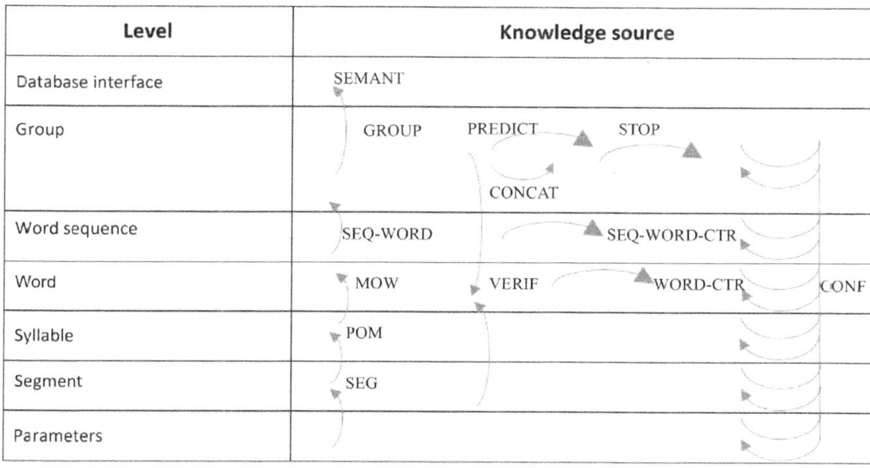

Figure 4.3. *Levels of knowledge in the Hearsay II blackboard*

One of the advantages of the blackboard model is that it facilitates the software integration of models written in different languages as well as the addition of a new module or the modification of an existing module. Despite these different advantages that make the model appealing for interactive systems, its conceptual complexity and the difficulty of implementing it remain major disadvantages. This complexity also affects the global performance of the system. From a cognitive point of view, a strong centralization of processes is likely to bias the emergence of collective behaviors. This pushed researchers to adopt architectures that have a similar general concept, but a simpler structure. These are hub-based architectures, which are central but passive data repositories. Several dialogue systems have adopted this architecture, including the CU Communicator of the University of Colorado [PEL 01, PEL 00] and the Vico multilingual dialogue system. Highly interactive, Vico is a human–machine dialogue system that serves as a vocal interface for a GPS equipped in a car, whose role is to guide the driver not only to find his route toward a given destination but also to provide tourist information and indicate service stations.

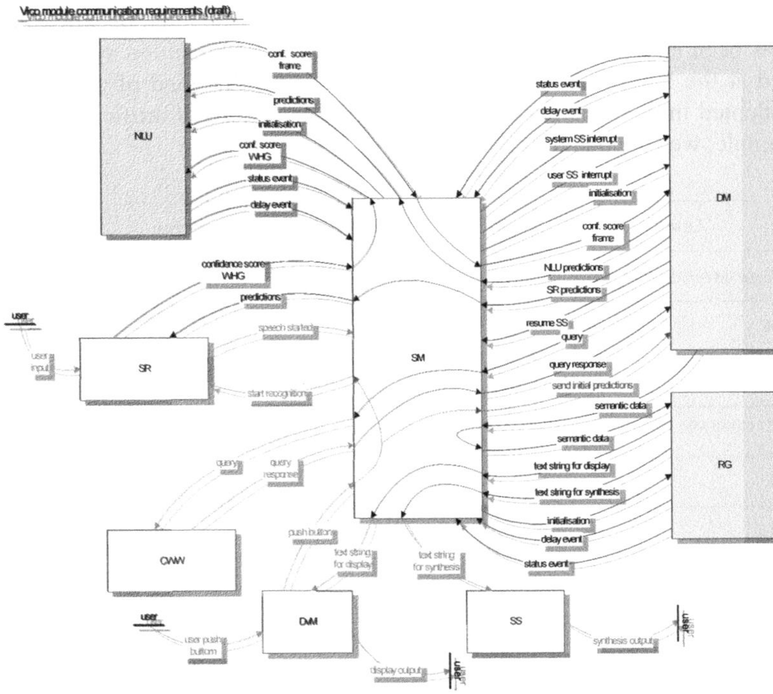

Figure 4.4. *Architecture of the Vico system [BER 02]. For a color version of this figure, see www.iste.co.uk/kurdi/language2.zip*

The dialogue manager plays a particularly important role within this system. As shown in the data flow diagram in Figure 4.4, this model interacts with the majority of the system components by providing delay queries and instructions for the four modules of linguistic processing: recognition, understanding, generation and synthesis, all while making requests on the Car Wide Web (CWW), the main source of information about the world in this system. The interaction of modules occurs through the CORBA[1] interfaces, a standard that was chosen in particular for the heterogeneity of the modules and the necessity of accessing CWW information through the network. The dialogue manager or hub is an obligatory passage for all inter-modular communication. Each module deposits its output there and searches for information that it could take as input there.

1 Common Object Request Broker Architecture is a conceptual framework and software that allows different applications to communicate with each other independent of their differences in terms of design or coding languages.

4.1.5. *Object-oriented architectures*

Object-oriented architectures are both a conceptual view of the breakdown of a program into units and a practical realization that guarantees the encapsulation of data and primitive operations within objects organized in a hierarchy of classes that facilitate the reuse of the code. This architecture has several advantages. From a common design, often made in the form of UML diagrams (class diagrams, collaboration diagrams, deployment diagrams, etc.), several programmers or even several teams can work in parallel on developing different classes of the same software, without risk of collision. The integration of new classes with an existing software or the modification of one or more existing classes is also facilitated. Contrary to models in serial architectures, the objects are not completely independent to the extent that each one must know the identity of the objects in its environment within the system and some of their properties, notably their input and output methods. This means that when an object is changed, all of the objects that invoke it must also be modified. In the NLP domain, object-oriented architectures are very commonly adopted, not only because of the advantages of this paradigm but also because of the popularity of many object-oriented languages.

One example of an NLP software developed with an object-oriented architecture is the GATE[2] (General Architecture for Text Engineering) system. Intended to process texts of all forms and sizes, GATE is an open-source software developed by a fairly large community. It was launched in 1994 by the Natural Language Processing Group at the University of Sheffield under the initiative of Yorick Wilks, Rob Gaizauskas and Hamish Cunningham. This system contains a large number of tools including an integrated development tool (GATE Developer) for NLP applications, which is widely used in the domain of information extraction; a collaborative annotation tool (GATE Teamwear) for the semantic annotation of data on an industrial scale and a process for the creation of robust services.

The GATE architecture is designed to take over the most important part of the engineering work to allow users to focus on developing their linguistic software. It also includes tools for syntactic analysis, labeling parts of discourse, information retrieval, information extraction, generating finite-state transducers for morphological and superficial syntactic analysis, evaluation and, most recently, interfacing with temporal expression extraction platforms [DER 16]. It is also

2 https://gate.ac.uk/

possible to integrate tools used in the domain of automatic learning such as Weka[3] [EIB 16] and SVM Lite[4].

4.1.6. *Multi-agent architectures*

The idea of multi-agent systems comes from a collective conception of intelligence that emerges from the behavior of a group [MIN 88]. Multi-agent architectures are an extension of multi-expert systems, blackboard-based architectures and object-oriented architectures. The advantage of these architectures is that they make it possible to simulate the behavior of several types of systems, including cognitive systems, social systems and systems that mimic animal behavior (ants, bees, fish, etc.). To learn more about multi-agent systems, see [WOO 95, BOI 01, CIA 01].

Let us begin with the definition of the concept of agent. In the literature, two definitions of agents are opposed: weak and strong. According to the weak definition, an agent is an electronic system or software that is characterized by the following properties: autonomy, social capacities, reactivity and proactivity. In other words, it consists of completely autonomous entities that have internal objectives and that submit to community or social rules of interactions with their peers. The strong conception of an agent stipulates that it has, in addition to the previously mentioned properties, anthropomorphic characteristics such as mental states, beliefs, intentions and emotions. In this case, the agent is often represented by an animated character or even a robot.

Agents are located in an environment made up of external elements that affect their behavior. There are two types of environment: physical environment and communicative environment. The physical environment provides the principles and the processes that govern and support the population of entities, and the communicative environment defines the principles and the infrastructure necessary for the transmission of information by the agents.

In the domain of NLP, the first forms of multi-agent systems were developed as multi-expert systems, following the example of the Caramel system, designed to promote collaboration between several expert systems specialized in processing a knowledge source [SAB 90]. Later, several multi-agent architectures were adopted for various goals. For example, the design of the architecture of the TALISMAN system intends to resolve linguistic ambiguities by making accessible all of the

3 http://www.cs.waikato.ac.nz/ml/weka/
4 http://svmlight.joachims.org/

necessary information for each linguistic level represented by the corresponding agent [STE 95]. The objective of the MICRO system was to propose a cognitively valid architecture [CAE 94]. Realized using the generic MAPS platform that is intended for the design of multi-agent systems, MICRO simulates the organization of the brain into two hemispheres: the analytic left hemisphere and the holistic right hemisphere (see Figure 4.5). The holistic processes are responsible for a global and synthetic Gestalt-type analysis, and the left hemisphere is considered to be the language center. Thus, two groups of agents distributed on two paths are distinguished: linguistic agents on the analytical and phonetic path and prosodic agents on the holistic path. The prosodic analysis conducts a rapid pre-labeling of events to guide the left path. The acoustic analysis at the input of the system is carried out using a mathematical model of an ear that has several processing stages [CAE 79].

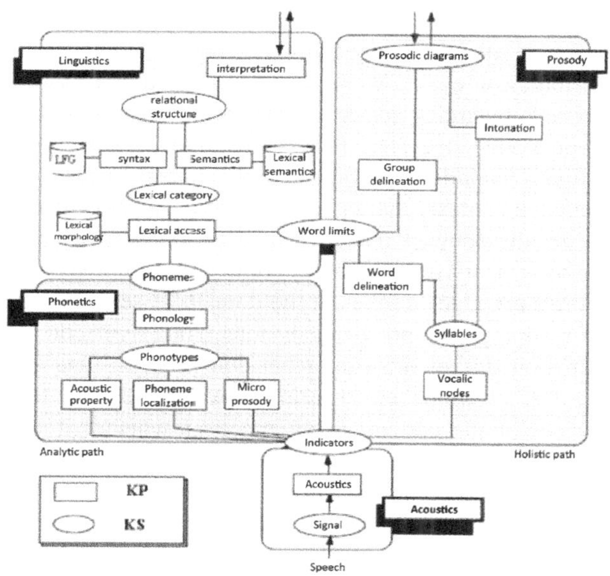

Figure 4.5. *Architecture of the MICRO system [CAE 94]*

Systems designed based on the strong conception of agents equipped with dialogic and emotional capacities have been developed [CHI 02, RIV 11]. Multi-agent systems have also served as a basis to experiment with fundamental hypotheses about the emergence of a form of linguistic code within an artificial community of agents. On this subject, the work of [KAP 00] focuses on the emergence of a lexicon within a population of artificial agents based on minimal hypotheses inspired by Wittgenstein's designation games.

4.1.7. *Syntactic–semantic cooperation: from cognitive models to software architecture*

The correlations between the architecture of an NLP system and cognitive models are very strong. To demonstrate this, consider the case of the interaction between syntax and semantics. This choice is motivated by both the interest in this question from theoretical linguistic and psycholinguistic perspectives as well as the central role played by syntactic parsing and semantic analysis modules in NLP applications.

It is a well-received idea in linguistics that semantic analysis occurs after the construction of a syntactic representation of an utterance. However, there is no consensus within the experimental psycholinguistics community regarding the mode of integrating syntactic and semantic knowledge. Some researchers argue that humans first proceed with the selection of an initial syntactic interpretation, which shows the priority of the syntactic analysis over semantic analysis [FOR 79]. Similarly, some studies on eye movement, including that of [RAY 92], have shown that the first decisions regarding sentence analysis, such as minimal attachment, are based on syntactic knowledge. The researchers showed subjects utterances with prepositional groups including ambiguous attachment. These utterances were anchored in discursive contexts that either did or did not support minimal syntactic attachment. The results showed that the subjects followed a garden path in the analysis and that the discursive context did not help to find a syntactic analysis. On the other hand, there are many studies in the domain of semantics that have shown the existence of a strategy for precociously integrating syntactic information with high-level information during the comprehension process. One example of these studies is [TYL 77], in which an experiment was conducted to verify whether semantics intervenes at the end of an utterance or whether it intervenes during processing alongside syntactic analysis. To do this, they used ambiguous adjective–verb pairs such as *landing planes* in utterances such as [4.1]. The results of this experiment show that semantics intervenes before the end of the utterance. When the word *planes* was followed by a context-appropriate word (such as *are* for [4.1a]), the response time was quicker than in cases where an inappropriate word was in the same place (such as *is* for [4.1a]). This shows that the subjects have a semantic preference, induced from general knowledge about the world as well as semantic analysis of the start of the utterance, which they apply in order to choose the most plausible syntactic analysis:

a) If you walk too near the runway, *landing planes*… [4.1]

b) You've been trained as a pilot, *landing planes*….

In addition, [TYL 77] used passive utterances with relative propositions to show that semantics as well as referential context can guide the choice of syntactic structure as in [4.2]:

a) The teachers taught by the Berliz method passed the test. [4.2]

b) The children taught by the Berliz method passed the test.

Utterances similar to [4.2b] were judged to be grammatical more frequently than utterances similar to [4.2a]. The semantic difference between the two utterances seems to be the reason for this difference in judgment, given that it is more probable that a child will be taught than a teacher. The researchers also showed that semantic cues intervene before the end of the utterance or even before a group boundary. Therefore, [CRA 85] concluded by considering the garden path as a contextual phenomenon, which can be avoided by knowledge of the context in which the utterance is produced. For example, according to the researchers, utterances such as [4.3a] were judged to be ungrammatical because their semantic bias distanced them from the hypothesis of a relative structure and led them to a *garden path*,[5] when they encountered the verb *passed*. This garden path was avoided in utterances similar to [4.3b], where the semantic context made it possible to guide the subjects toward a relative structure that was syntactically correct. A similar study to that of Crain and Steedman was conducted by [TRU 92] from a discourse analysis perspective. The researchers used utterances with ambiguous verbs (whose form is identical to equivalent past participles) as in [4.3]:

a) The fossil examined.... [4.3]

b) The archeologist examined....

In these utterances, the verb *examined* has the same form as the past participle of the same verb. As in [CRA 85], the researchers found that semantic knowledge directly influenced the choice of syntactic structure and sometimes led to garden paths.

[CAR 88] conducted a study of eye movement during reading. This work showed that the duration of fixation on words that were semantically abnormal to the context was longer than that of equivalent words (in terms of length, frequency and syntactic adaptation compared to the rest of the utterance) that were semantically normal. This shows that the semantic analysis occurs in parallel with the reading process (and therefore with the syntactic analysis). Several experiments aiming to clarify the role of the discursive context in comprehension have been

5 A *garden path* is a case of local ambiguity that guides the human processor toward a single analysis from which it is difficult or sometimes impossible to make a correction in the analysis (see [CRO 96] for more details about this phenomenon).

conducted [SPI 93, BOL 95]. In his study, [ALT 99] described two experiments concerning this problem. In these two experiments, the subjects had to read utterances similar to [4.4]. These utterances were used in contexts that introduced possible objects. The idea was that after the verb *drank*, the subjects were supposed to think that the utterance did not make sense if the object of the verb was not a drinkable element. After asking the subjects to examine different groups of utterances, he observed that the negative answers (i.e. that the utterance did not make sense) required more time when the anterior context was ambiguous. The author concluded that the semantic roles (agent, patient, recipient, etc.) associated with arguments that are discursively anterior to a verb (arguments located in a prior conversational turn) are selected at the moment of perceiving the verbal head by considering the available roles (i.e. roles that have not yet been associated with a lexical unit), even when the entity that refers explicitly to these antecedents (anaphoric pronouns) is in a post-verbal position and this entity has not yet been processed by the subjects. This shows that the discursive context intervenes in the syntactic analysis of an utterance and that this intervention occurs in parallel with the syntactic analysis, given that its effect is detected before the end of processing the utterance:

He drank some.... [4.4]

The disagreement between the results of these experimental studies can be interpreted in two different ways. On the one hand, it could be the result of methodological or technological issues in the investigation of the considered problems. On the other hand, it could be the reflection of the existence of a variety of strategies whose triggering is produced by specific conditions. This divergence in the cognitive domain has affected the divergence of the architectures of NLP systems that are divided into three main forms: sequential, integrated and parallel [MAH 95] (see examples in Figure 4.6). The architectures (a) and (b) in Figure 4.6 are relatively simple to explain because they are the direct translation of syntax first or semantics first strategies, respectively. They presuppose the existence of independent knowledge sources and independent processing methods for both syntax and semantics. The interaction between the modules occurs unidirectionally from the source toward the destination. The integrated architectures (c and d in Figure 4.6) in turn start from the premise that there is a unified framework for the representation of syntactic and semantic knowledge as well as for their processing. In practice, the use of modules based on rich formalisms such as HPSG, which integrates morphological, syntactical and semantic knowledge, is an example of an integrated architecture.

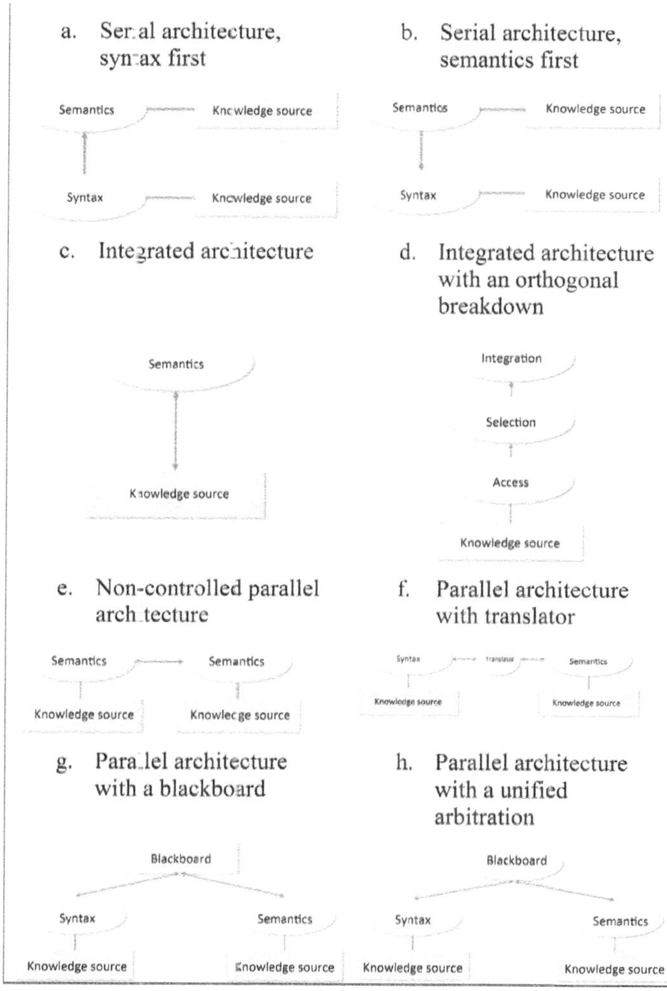

Figure 4.6. *Some software configurations for integrating syntax and semantics*

More complex, parallel approaches use independent knowledge sources and processing that occurs on two axes simultaneously. These can be realized in more varied forms than sequential or integrated approaches, as in the four forms, (e), (f), (g) and (h), shown in Figure 4.6. The main reason for this divergence is the mode of coordination between the two parallel axes. Some architectures implement a blackboard-based coordination whose role is to coordinate efforts, while others implement an arbitration that attributes a fixed or context-dependent weight to resolve potential conflict between the two knowledge sources. Non-controlled parallel architectures presuppose the adoption of a single generic framework for syntax and semantics that makes direct communication impossible. When the two modules have different theoretical backgrounds, a module to translate formats or convert content is required for all interactions between these two modules.

4.1.8. *Programming languages for NLP*

Although NLP applications can be coded with a very large number of high-level programming languages, notably including Java, C++, C# and Python, some seem more appropriate for language processing due to either their inherent properties or the massive adoption of these languages by the community and, consequently, the availability of rapid development tools for NLP written in these languages.

Historically, LISP, designed by John McCarthy in 1958, is the programming language that first attracted the interest of NLP experts, notably because of the ability to process lists and lambda calculus (see [GAZ 89a] for an introduction to NLP with LISP).

For a long time, Prolog was very directly linked to NLP. In fact, this language was designed with the intention of automatically processing language (Colmerauer's metamorphosis grammars) and Definite Clause Grammar (DCG). This language is an implementation of predicate logic that offers a privileged framework for semantic representation. The native integration of backtracking considerably facilitates the implementation of a large number of NLP algorithms, mainly including parsing algorithms. Another important advantage of Prolog is the availability of books that cover the main topics of NLP such as [PER 87, GAZ 89b, MIC 91, COV 94]. Two major families of Prolog can be distinguished, Marseille Prolog or Prolog II and Edinburgh Prolog, which is more common. Following the interest in this language, several implementations of the Edinburgh version were created, each with specific properties. Swi-Prolog is an academic version that is largely used for teaching around the world. Developed in the Netherlands at the University of Amsterdam, this version of Prolog is distinguished by respecting the Edinburgh syntax as well as is efficiency. It is also distributed with an open-source license. SICStus Prolog was developed at the Swedish Institute for Computer Science and is also popular in academic environments. Gnu Prolog is a free version developed at the INRIA. Amzi-Prolog is distinguished by its easy integration with Java, notably in the context of the Integrated Development Environment (IDE) ECLIPSE. A comparative study of the performances of several versions of Prolog showed that there are substantial differences in performance between these different versions (see Table 4.1).

System	Version	Time
SWI	1.8	1.2
SICStus	3.11.2	11.8
Yap	4.4.4	19.5
Ciao	1.10	6.6
GNU	1.2.16	10.9

Table 4.1. *Comparison of several programming environments in Prolog [FER 00]*

Despite its many advantages, Prolog also has several disadvantages. Based on logic, Prolog syntax is difficult to learn compared to other languages with more conventional forms for computer scientists. The many variants also make it difficult to interface with programs written in other languages. Although it is very well suited for processing logical formulas, Prolog is not a natural tool for implementing statistical algorithms or matrix calculations, which are very important for many NLP applications. The most significant disadvantage of Prolog is probably its low speed compared to other lower-level languages such as C, Java or Python.

Python is increasingly popular within the NLP community. Developed in 1991 in the Netherlands by Guido Van Rossoum, this language is included in the list of languages officially adopted by Google. The advantage of Python is that it can be used as an object-oriented language or as a script language whose syntax is intended to have a simplicity similar to natural language. This gives it the advantage of greater generality compared to a strictly object-oriented language such as Java with slightly more complex and strict syntax. In other words, Python can be used to write small projects quickly, large commercial applications and a glue code to integrate programs written with languages such as C, C++ and Java. Python is a particularly attractive language for developing Web applications, notably thanks to the availability of libraries or frameworks like Zope, CherryPy or Django. Python also has rich documentation as well as abundant literature that covers practically all aspects of programming related to this language [LUT 06, DAW 03, ASC 03, MER 03].

For an NLP specialist, what make Python an attractive language are the data structures such as lists and dictionaries that facilitate the processing of character sequences. External tools developed with Python make it even more attractive, including the NLTK toolbox, which includes morphological, syntactic and semantic analysis tools. Particularly easy to use and well documented, it is used in research as a support for practical works in NLP courses [BIR 09] and in the industry. In Python, the availability of other tools and generic libraries related to the domains of artificial intelligence and machine learning that also apply to NLP, for example Orange[6], Scikit Learn[7], PyBrain[8], PyML[9], LibSvm[10] and Neurolab[11], make this language even more attractive. The major disadvantage of Python is its relative slowness compared to languages like C. As an indication, an informal comparison of the speed of loop processing was conducted between three languages: C (reputedly the most efficient of the high-level languages) with the MinGW compiler (a minimalist version of the GNU compiler for Windows), Python (2.7) and Prolog (Swi-Prolog). Realized on a machine with a Pentium III processor and Windows 2000, the experiment had the following results: the code written in C was 28 times faster than the code written in Python, and Python was, in turn, 10 times faster than the code in Prolog.

4.1.9. *Evaluation of NLP systems*

The question that is always posed when developing an NLP software is: does this software respond to the prerequisites initially expressed in the project specifications? The evaluation is an objective way to answer this question. There are two types of evaluation, depending on when the operation is conducted: formative evaluation and summative evaluation. Formative evaluations occur during the development of the program on prototypes or intermediary versions. The objective of these evaluations is to inform people involved in the development, of possible problems regarding the system design or its code. Summative evaluations are done on an intermediary or final version of the system and are intended to evaluate the system's performance and the degree to which it satisfies the project specifications. A comparison of these performances with the performances of one or more similar systems can be one of the objectives of a summative evaluation. The evaluation can be done manually or automatically using a battery of tests.

Regarding the object of the evaluation, two different dimensions of the software are verified: structural dimension and functional dimension. The structural dimension concerns the code and its internal logic in a way that is almost independent from the system requirements. The idea is that software with errors or bugs in the code cannot satisfy the specifications, whatever they may be. The functional evaluation, in turn, is intimately linked with the system requirements. The results of this evaluation are a good indication with which to judge the quality of the cognitive or linguistic model adopted and sometimes the quality of the data used in

6 https://orange.biolab.si/
7 http://scikit-learn.org/stable/
8 http://pybrain.org/
9 http://pyml.sourceforge.net/
10 https://www.csie.ntu.edu.tw/~cjlin/libsvm/
11 https://pythonhosted.org/neurolab/

the development process. There are two types of functional evaluation: quantitative and qualitative. For a general introduction to the evaluation of NLP systems, see [RES 10, PAR 07, KIN 98]. For more information on the structural test as well as other forms of similar evaluation, see [LEW 05].

4.1.9.1. *Structural test*

Sometimes called the *whitebox test*, the structural test is intended to evaluate a program's logical paths. This kind of test pertains more to the internal logic of the program than to the specific requirements in the project specifications. It requires very deep knowledge of the internal structure of the program and its logic. It is intended for the identification of malfunctions or bugs in the code or malfunctions related to module compatibility. In heterogeneous software systems, partial incompatibility of module input and output formats is a major source of errors. A complementary test, not for errors but for the runtime speed, can reveal problems in the implementation. The idea is to prepare a battery of input sentences of varying lengths and measure the runtime for each one. This makes it possible to draw up a line of the runtimes. Comparing the practical curve with the theoretical complexity of the algorithm implemented makes it possible to identify the problems. For example, to evaluate the runtime of the Oasis system, a corpus of 588 utterances was chosen [KUR 03]. The utterances chosen were extracted from the hotel reservation corpus and the corpus collected in the challenge evaluation campaign which will be presented in section 4.1.9.3. To get a clearer idea of the runtime curve, consider the graph that only contains the worst runtimes observed (Figure 4.7). The observation of all of the utterances was confirmed with the analysis of these cases that generally show a linear behavior. Note that the exceptions to the gradual increase in runtime with the increase in utterance length are due to the difference in frequency terms between the different lengths.

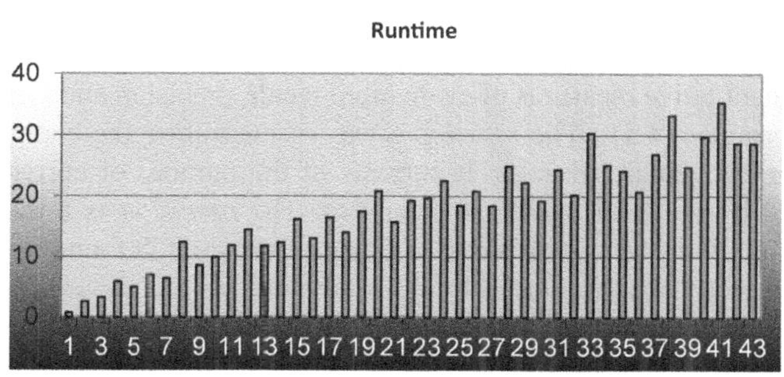

Figure 4.7. *Runtime of the worst cases observed by length*

4.1.9.2. *Quantitative evaluation*

This type of evaluation is intended to establish a general idea of a system's performance. It is done using a set of representative cases that is sometimes called a battery of tests. Typically, the corpus is divided into two unequal parts. The first part, generally two-thirds of the cases, is used for training the system or writing the rules. The second part, composed of the rest of the cases, is used for the evaluation. The motivation behind this separation is the evaluation of the generalization of the system, which is related to its capacity to process new cases. In practice, this division does not guarantee a complete diagnostic of the limits of a system because it depends on the variation of the data in the corpus (linguistic structures, vocabulary, etc.) but nevertheless remains a minimum. Note that the evaluation can also concern an independent module (known as a modular evaluation) rather than the entire software.

When evaluating any system, there are four types of cases:

– True Positives TP: acceptable cases that are considered as such by the system. For example, in the case of the evaluation of a syntactic parsing module, an example of this type would be a well-formed sentence that is correctly parsed by the system.

– True Negatives TN: unacceptable cases that are rejected by the system. In the case of a syntactic parsing, an example of this type would be a malformed sentence correctly rejected by the system.

– False Positives FP: unacceptable cases that are incorrectly analyzed as being acceptable. This is sometimes referred to as overgeneration. This is an ungrammatical sentence that is incorrectly syntactically parsed as a correct sentence.

– False Negatives FN: correct cases that are incorrectly rejected by the system. For example, a grammatical sentence incorrectly rejected by the parser.

Having established a typology of the analysis cases, the next step is to define the three main quantitative measures of evaluation: recall, precision and F-measure or F-score (see formulae [4.5]). The *recall* is a way of measuring the completeness of a system or its horizontal coverage. It consists of the number of correctly analyzed cases divided by the total number of analyses. The *precision* is a measure of the processing quality or the system's rate of overgeneration. It is found by dividing the number of correctly analyzed cases by the total number of analyses. The idea is that a system that relaxes its constraints results in good recall but that this causes a decrease in the precision. Inversely, if the system is too restrictive, the precision will be good but the recall will be low. A good NLP system must find the right balance between the recall and precision. To unify these two measures, Nancy Chinchor [CHI 92] proposed the F-score:

$$\textbf{Recall} \quad \frac{TP}{TP + FN} \qquad \textbf{Precision} \quad \frac{TP}{TP + FP} \qquad \textbf{F-score} \quad \frac{2 \times \text{precision} \times \text{recall}}{\text{precision} + \text{recall}} \qquad [4.5]$$

Evaluating one's own system is often a biased process despite the good intentions of the system's designer. Besides using test data from the same corpus as the training, an evaluation by the designer risks repeating use scenarios that are typical for the system and avoiding scenarios that could mislead the system. It is also practically impossible to compare the performances of two different systems whose evaluation has been made using two different corpora. One solution to this problem is to proceed with collective evaluation campaigns of similar systems (with the same requirements). Especially in the competitive American tradition, this results in a sequencing of the systems involved in the campaign according to the results. Several evaluation campaigns have been organized first in the United States and then in Europe. For example, there have been campaigns to evaluate TREC information retrieval systems [VOO 05], tagging parts of speech [ADD 98], understanding spoken or written language with the ATIS and MUC campaigns [CHI 92] and analyzing speech [ANT 02b].

4.1.9.3. *Qualitative evaluation*

A qualitative evaluation method should satisfy a set of requirements. For example, it should be generic and independent of all applicative domains. It should also be applicable to systems whose output is encoded in different formats. It should also provide a detailed diagnostic of the system's behavior that exposes its weaknesses and strong points. This diagnosis should ideally include all relevant phenomena while having the possibility of establishing a partial diagnosis per phenomenon. For example, if we wish to evaluate a syntactic parsing module, it must be possible to evaluate it on all syntactic phenomena such as subordination, coordination and displacement, but a partial evaluation on a system on a limited number of phenomena such as negation or coordination should also be possible.

To gain a deeper understanding of a system's behavior, not only can the system's recall and precision be given for the entire test corpus, but these measures can also be used per phenomenon of interest. For a syntactic analysis module, the recall and the precision can be found for sentences with ellipses, coordination, subordination, etc. This makes it possible to establish a deeper diagnosis of the system's performances. It is also sometimes useful to draw up a record of the kinds of errors in terms of confusion. In particular, this makes it possible to determine the percentage of confusion between the different phenomena. To do this, a confusion matrix can be established in the form of a table where each cell [i, j] indicates the prediction frequency of a label *j* in the place of the label *i*.

The DCR method (Declaration Control Reference) is an attempt to satisfy the requirements of an objective evaluation. The basic idea behind this method is to produce a derived corpus that can be used to evaluate the system from an initial corpus, based upon which the system was constructed. The derived corpus is composed of several parts, each of which is dedicated to evaluating a specific linguistic phenomenon. On the one hand, this is a way of bypassing a lack of relevant data and, on the other hand, this makes it possible to guarantee the objectivity of the evaluation. A DCR evaluation has three steps [ANT 00]:

1) *The declaration D*: consists of an ordinary utterance that features in the system's training corpus.

2) *The control C*: consists of modifying the version *D* of the utterance in order to account for a specific phenomenon.

3) *The reference R*: is a Boolean value that accounts for the coherence of the utterance.

To clarify these three steps, consider this example:

<D> I would like a double room with a view of the sea.

<C> I would like a single room.

<R> False.

In order to make the process of deriving C utterances from D utterances more systematic, a modified version of DCR, called extended DCR, was proposed by [KUR 02]. The primary innovation consists of using a description, the most exhaustive possible, of the intended phenomena called the derivation grammar as the basis for the derivation process.

A combination of objective evaluation methods and collective evaluation campaigns was also proposed in [ANT 02b]. This campaign, called the challenge evaluation, involved five automatic spoken language understanding systems from four French laboratories specializing in NLP and human–machine dialogue systems. These include the CLIPS-IMAG laboratory in Grenoble, the IRIT laboratory in Toulouse, the Valoria laboratory in Vannes and the LIMSI laboratory in Paris. Each system designer proposed his own initial corpus composed of 20 initial utterances that his system was capable of processing correctly. Proposed to the project partners, the initial corpus served as a basis on which to obtain the derived corpus. Having the only constraint of not leaving the semantic domain of the initial utterance, the partners were strongly encouraged to use their imaginations to propose original structural modifications. The modifications could have the form of grammatical phenomena (extractions, parentheses, ellipses, etc.), extra-grammatical phenomena (repetitions, hesitations, etc.) or simulations of artificial phenomena such as recognition errors (the subjects remove, replace or add words in a similar way to what recognition systems can do in cases of error). For each initial utterance, each participant created 15 derived utterances. In other words, for each initial utterance, there were 60 derived utterances and 1,200 utterances in the derived corpus. As an example, a translation of an initial utterance and a set of derived utterances are given in Figure 4.8. Before being used, the derived corpus was validated by the designer of each system. The validation consisted of a judgment by the system's creator of the adaptation of the derived utterances proposed by the campaign participants. Utterances deemed unsuitable were modified by the creator of the test. The main requests that were made by the participants concerned spelling errors as well as utterances judged not relevant or not realistic in relation to the system's task.

Initial utterance:
<1> alright in that case then reserve me a nice quiet room for next February 26 </1>
Five derived utterances generated by the IRIT designer:
<1.1> alright in that case then reserve me a nice and um quiet room for next February 26 </1.1>
<1.2> alright in that case then reserve me a nice quiet room for um next February 26 </1.2>
<1.3> alright in that case then reserve me a room um a nice quiet room for next February 26 </1.3>
<1.4> alright in that case then reserve me a nice quiet room for next February 25 um no that's not it next February 26 </1.4>
<1.5> alright in that case then reserve me a nice quiet room for next February 25 um 26 </1.5>

Figure 4.8. *An example of an initial utterance and some derived utterances*

4.2. Machine translation (MT)

Today, globalization has made enormous quantities of information accessible, for example, in the form of web documents or texts scanned and transcribed with OCR[12] software, but use of this information sometimes faces a linguistic barrier. New multi-national organizations are being added to many existing organizations such as the World Trade Organization (WTO) and the European Union. This need has grown even more with the arrival on the global stage of certain Asian countries whose

12 OCR (Optical Character Recognition) makes it possible transform the letters in texts scanned in image form into computer characters similar to those entered in the program Microsoft Word.

languages are little-known outside their national borders. It should also be noted that, far from its ideals, MT is an excellent means of technical and industrial spying. Human translation, the natural response to all these needs, suffers from several disadvantages, notably its high cost and the need for a foreign human presence which is not necessarily possible or desirable depending on location (translation of military texts or conversations, for example), the time of day or year or the subject in question. Hence, there is a need for a software system that is capable of taking the input of a text in a source language and producing an equivalent text in a target language. The advantage of these systems compared to human translation is the spatial and temporal availability (a telephone or internet connection is sufficient to access the system when it is deployed in the form of Software As a Service (SaaS)). For a general introduction to machine translation, see [HUT 92, ARN 94, ACL 16].

4.2.1. *Why is translation difficult?*

Translation is a very difficult task and, whatever the skill level of the human translator, translation errors are always possible. The constantly changing dynamics of language is a common reason for these errors. In addition, some texts require extra-linguistic knowledge related to the content that the translator does not necessarily have. This contributes to an incomplete or even incorrect understanding of the original text and, consequently, a bad translation.

4.2.1.1. *Linguistic divergences*

From a lexical point of view, cases of polysemy where a single word can have several different meanings are probably the most important. For example, the verb *know* can be translated as *savoir* or *connaître* in French depending on the context (see examples [4.6]).

1) I know the book you're talking about.	3) Je **connais** le livre dont tu parles.	
		[4.6]
2) I know how to go to England.	4) Je **sais** comment aller en Angleterre.	

In addition, certain cases of homonymy can also be a source of problems. For example, the British words *pound* (unit of weight and currency) are impossible to distinguish. Sometimes the context makes it possible to disambiguate the meaning but not always. Similarly, the French word *langue* is one example that linguists know well. This word does not have a direct equivalent in English, which is why English linguists are obliged to borrow the French word for their texts, notably when they are discussing the famous Saussurian distinction of langue/parole. Many words or expressions do not have an equivalent in other languages. For example, the notion of the *weekend* does not exist in the Arab world in the same way as in the Western world. Consequently, there is no clear equivalent in Arabic of the French word *weekend*, which is itself borrowed from English[13]. Moreover, the locutions *right away*, *see you soon* and *see you in half an hour* do not have an exact equivalent in Arabic. Obviously, a more or less equivalent expression can always be found; however, as the concept itself is not familiar, this expression will seem unexpected or even bizarre.

Given the syntactic differences, the correspondence between two sentences belonging to two different languages, regardless of their degree of similarity, seems impossible. Word order differs from one language to the next. For example, French and English are languages where the preferred sentence order is Subject Verb Object (SVO), German is a SOV language and Arabic is a VSO language. Similarly, agreement does not have the same form in all languages. In French, for example, the verb agrees with its subject in number but not in gender, unlike Arabic, where the verb agrees in number and gender. Note that there are also particular forms like the subject pronoun that are not given in sentences in Italian (see [4.7]):

Parlo bene l'italiano
I speak Italian well. [4.7]

The binary in Arabic, a particular form of the plural that, to my knowledge, does not have an equivalent in European languages, is another example of these particular cases.

4.2.1.2. *Ambiguities in the source text*

Ambiguity is one of the key problems in NLP. Ambiguities concern all linguistic levels from the words to the text. Resolving these ambiguities is not a simple task. Even humans with a good understanding of the context sometimes have difficulties.

4.2.1.3. *Errors in the source text*

In real applications, notably on the Internet, the level of formality and therefore of grammaticality of texts can vary substantially, which can lead to reduced performance of machine translation which makes manual correction necessary.

13 The weekend in Arabic literally means the last days of the week, which are not all holidays. In some countries, Friday is the only holiday at the end of the week, while in others, it is Thursday and Friday and in others, it is Saturday and Sunday.

4.2.2. *History of MT systems*

4.2.2.1. *The early days: (late 1940s–1965)*

The beginnings of machine translation go back to the late 1940s, with the work of W. Weaver among others, who proposed using cryptographic decoding techniques to translate texts automatically [WEA 55]. In 1952, the first conference on machine translation was held at the Massachusetts Institute of Technology (MIT). This conference occurred in the context of the American–Soviet rivalry during the Cold War. The first prototype called the Georgetown Automatic Translation (GAT) was created at Georgetown University by the team of D. Dorstert and R. McDonald. The hopes elicited by MT systems were quickly diminished. These systems, called the first generation or direct translation, were based on a word-for-word translation model that only made recourse to a dictionary, without using syntactic or semantic representations. Critics of the systems during this period continuously emphasized the errors that these systems produced during the translation of two English proverbs into Russian and back into English, to ridicule MT (see, for example, [4.8] for a French adaptation):

The spirit is willing but the flesh is weak	La vodka est bonne mais la viande est pourrie. *The vodka is good but the meat is rotten.*	[4.8]
Out of sight, out of mind	Invisible et fou. *Invisible and insane*	

At the same time as the work conducted in the United States, Soviet researchers focused more on theoretical and linguistic subjects. Already at that time, their work considered the possibility of creating a universal formal language that could serve as a bridge between the source language and the target language.

Following the controversy over the performances and use of MT systems, the American government agencies who had funded the research work of about 20 teams in MT created a special commission called the Automatic Language Processing Advisory Committee (ALPAC) in order to evaluate the progress made by the teams in question. Under the direction of John R. Pierce, this committee wrote a report that can be summarized by the following points [ALP 66]. First, the mediocre quality of the translation obtained by these systems often required human intervention to obtain a good translation. Next, because microcomputing was non-existent at the time, the development of such systems was particularly expensive. Finally, it seemed necessary to consider the linguistic factors and consequently increase support for computational linguistics. However, it was the negative content that the readers of this report retained and it resulted in a decrease in funding for MT projects and MT research was set on the back burner for several years.

4.2.2.2. *The second generation (1965–1975)*

During this period, despite the decline in funding, some teams nevertheless managed to continue their work on MT systems, including:

– the METAL system (University of Texas);

– the SYSTRAN system;

– the TAUM system (University of Montreal);

– the CETA/GETA system in Grenoble directed by Bernard Vauquois.

Most of the systems in this generation adopted an "indirect" approach where intermediary semantic representations independent of language served as a point of passage between the source language and the target language. From an architectural point of view, these systems opted for a strict separation between linguistic resources, notably syntax and semantics on the one hand, and software resources on the other hand. Aiming for better translation than the first-generation systems was also an objective of this generation of systems, which were characterized by these features:

– The syntactic analysis was gradually enriched by a semantic and even contextual analysis.

– The domain of application was restricted, for example, to the translation of weather reports, like the TAUM-METEO system from the University of Montreal.

– The vocabulary and syntactic constructions were relatively small in number and therefore easily catalogued and the risks of ambiguities decreased.

– The languages covered diversified and expanded beyond the Russian–English pair.

4.2.2.3. *Resurgence in interest (1975–1990)*

After 1975, there was an increase in the need for translation systems by companies. This resurgence in interest was also supported by the development of new computer and electronic tools, such as programming languages and modeling, and the availability of more powerful computers. The systems developed at this time were mainly the most advanced and most sophisticated second-generation systems. The pace of work on MT accelerated in the early 1980s, as demonstrated by these projects:

– the PN-TAO project between 1983 and 1987 (France);

– the EUROTRA project launched by the European Commission in 1982 that covered its nine majority languages;

– the national Canadian program launched in 1985;

– the programs of fifth- and sixth-generation systems launched by the University of Kyoto in Japan;

– the SYSTRAN software was adapted for MS-DOS and ran on the IBM PC in its version with two disk drives, meaning without a hard drive.

With developments in linguistics, notably in syntax, several researchers explored the adoption of new linguistic formalisms based on unification or constraints in machine translation. The advantage of these formalisms was their capacity to incorporate several levels of representation with a single formalism.

4.2.2.4. *The recent period*

The power of computers made it possible to imagine a different solution: drawing from immense corpora of computerized databases to reuse fragments of already-translated sentences. Notably initiated by IBM, the approach grew in the 2000s with Google, which collects all the translations on the Internet, marking a turning point in the history of MT. This period has also seen the launch of initiatives to create large-scale linguistic resources. It is also distinguished by the creation of collection and distribution centers of resources such as the Linguistic Data Consortium (LDC) (University of Pennsylvania, USA) and the European Language Resource Agency (ELRA) (Paris, France). It has also seen the launch of the first conference dedicated to NLP resources, the Language Resources and Evaluation Conferences (LREC) (see Chapter 1 of the first volume of this book [KUR 16] for more details).

4.2.3. *Typology of MT systems*

The number of languages covered by an MT system is an important criterion when drawing up a typology of MT systems. From this perspective, two types of systems can be distinguished: bilingual systems and multilingual systems. Bilingual systems are divided into two groups: reversible or bidirectional systems, where translation can occur between the source language and the target language and vice versa, and unidirectional systems. The number of languages plays a very important role in the choice of technique to adopt for the translation.

Beyond the purely quantitative aspects, the number of languages concerned by a system affects the software aspects, the approach adopted for the translation. Another criterion for classifying an MT system is the input media. In fact, the input of an MT system can be a scanned text that is transcribed automatically by an OCR software. With the progress with OCR, this task is doable with the current technology except that some transcription errors can occur especially when dealing with a handwritten text. Other types of MT systems are designed to deal with spoken conversations. In this case, they should not only handle speech recognition errors but also the specific phenomena that can occur within spoken language such as extra-grammaticality. It should be noted here that prosody is not usually accounted for by MT systems.

Regarding the sharing of tasks between the machine and the human translator, these types can be distinguished: human translation, machine translation assisted by humans and machine translation.

In the case of human translation, all of the translation steps are completed manually. The only possible assistance by machine in this case is the level of formatting of the translation (word-processing software) or the level of access to lexical information (electronic dictionary or translation memories). It is important to note that purely human translation does not preclude the intervention of other human experts during or after the translation. This intervention is sometimes formalized by the addition of the name of the translator and the reviser in the meta-data of the translation.

Moreover, metalinguistic information, sometimes necessary to translate a text well, is also accessible through electronic encyclopedias or on the Internet. In some cases, a translator can use computer resources, which is known as Computer-Assisted Translation (CAT).

Regarding automatic translation assisted by humans, the input text can be modified by a human at three different steps. The first type of intervention can occur before the intervention of the machine, when a translator quickly reads through the text to correct potential spelling or grammar errors. Depending on the system, lexical processing can sometimes be carried out, for instance the replacement of words outside of the system's vocabulary. A first translation of the text toward a controlled natural language can also be done. In this type of language, the vocabulary and the syntactic constructions are limited in advance to control the ambiguities. The second type of intervention occurs during translation by the machine. Some MT software interacts with the user through dialogue to resolve certain ambiguities that the software could not resolve on its own. Requiring an alert user, this process becomes cumbersome when the text contains many ambiguities because the translation time increases considerably. Finally, the third type of intervention occurs after the machine intervention in post-processing to correct errors or improve style. Some MT software produce several alternative translations in cases of structural or semantic ambiguities, giving the translator the responsibility of choosing which one should be retained during this last step.

Machine Translation itself is an entirely automated process where the machine carries out all steps of the translation without any external intervention.

4.2.4. *The use of MT*

From a functional point of view, J. Carbonell established the distinction between two processes: assimilation and dissemination (see Figure 4.9).

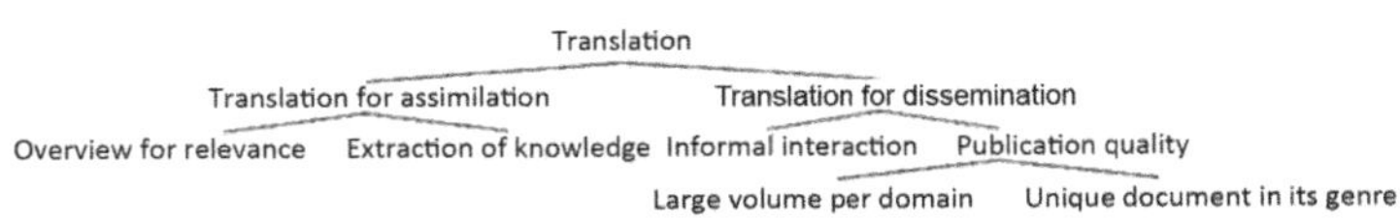

Figure 4.9. *Functional typology of MT systems according to Carbonell*

Assimilation consists of formulating a synthesis of what is written about a subject like a political, economic or sporting event in different languages (Figure 4.10) from a set of documents that often come from a collection on the Internet. Given the wide variety of topics that can be addressed, this cannot involve high-level knowledge, notably semantic and pragmatic knowledge, in the translation system. Consequently, it is expected that the quality of the translation will require human post-editing.

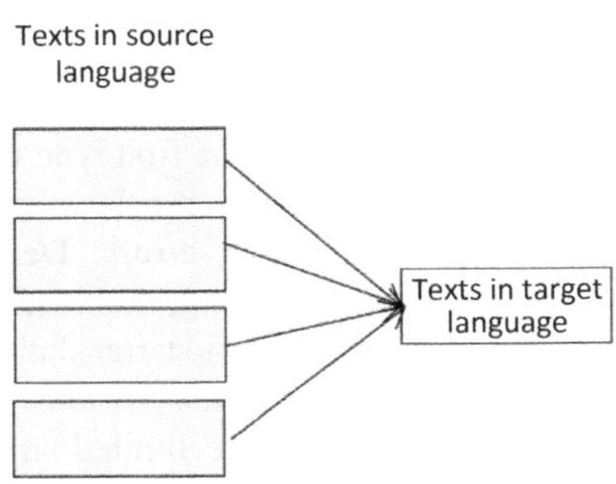

Figure 4.10. *General diagram of the assimilation process*

Dissemination, on the other hand, occurs from a given language toward several languages (see diagram in Figure 4.11). A typical example of this process is the translation of product catalogues intended to be exported. In general, this consists of translations that concern a limited domain. This makes it possible to carry out deep linguistic analyses and therefore allows for a better-quality translation that requires little or no post-editing.

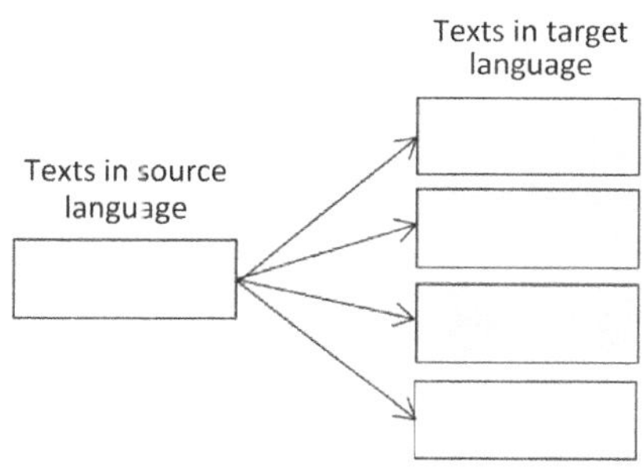

Figure 4.11. *General diagram of the dissemination process*

The translation of literary works, especially poetry, is distinguished from the two abovementioned types by the fact that the form and poetic function must be transposed. This type of translation requires very advanced skill in the target language. Generally, translators of this kind of literature are themselves bilingual authors or poets. In some cases, the translation is done by two people: a translator provides a translation of the meaning into the target language and an editor reformulates it in the target language in an appropriate literary style. For example, the most famous Arabic translation of *Paul et Virginie*[14] by Bernardin de Saint-Pierre was formulated by Egyptian author Mustapha-Lutfi al-Manfalouti, who did not speak French. An MT software can thus be used to obtain drafts whose final formulation will be done by a professional in the target language without necessarily needing to have perfect mastery of the source language.

4.2.5. *MT techniques*

It is possible to carry out a translation using several levels of linguistic knowledge. The more diverse the knowledge, the better the quality of the translation, but the more complex the development process. To represent this diversity of systems, Bernard Vauquois proposed the triangle given in Figure 4.12 [VAU 68].

14 https://en.wikipedia.org/wiki/Paul_et_Virginie

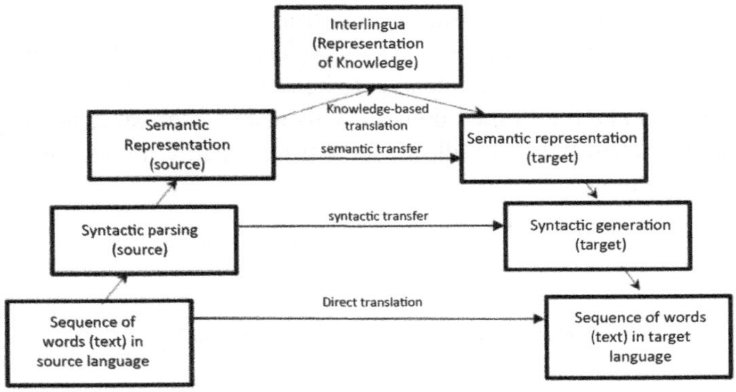

Figure 4.12. *The Vauquois triangle with some modifications [VAU 68]*

Four different approaches can be distinguished from the triangle: the direct approach, the transfer approach (syntactic and semantic) and the pivot approach.

4.2.5.1. *Direct approach*

Adopted by the majority of the first-generation systems, the direct approach is the most intuitive and the simplest. The translation process consists of finding an equivalent in the target language for each word in the source sentence to form the target sentence (see Figure 4.13). First of all, a superficial morphological analysis phase makes it possible to find the basic forms of words: lemmas. They are then used to find an equivalent in the target language with a bilingual dictionary, without any consideration for syntax or semantics. Finally, some local rules make it possible to order the words obtained, for example, by moving the adjectives or prepositions.

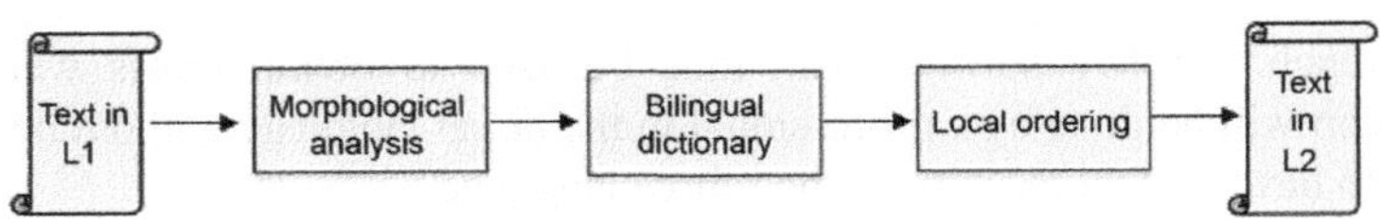

Figure 4.13. *Architecture of a system using the direct approach*

Because it does not consider the context, this approach comes up against all kinds of ambiguities. Moreover, the local rules for re-ordering are not sufficient and sentences obtained in the target language are often not correctly formulated.

4.2.5.2. *Transfer approach*

These systems are less ambitious than the pivot systems, because they provide two distinct representations *R* and *R'* whose level of abstraction is lower than the pivot language (see Figure 4.14).

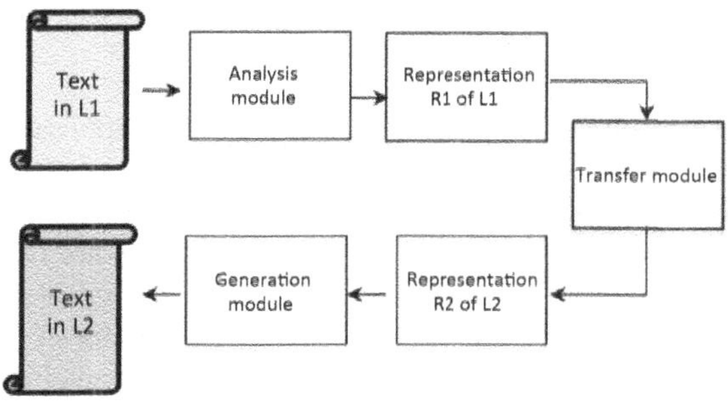

Figure 4.14. *Architecture of the transfer approach. For a color version of this figure, see www.iste.co.uk/kurdi/language2.zip*

The targeted representations are often syntactic–semantic constructions of sentences instantiated by words in the language. They are therefore concrete linguistic representations rather than abstract conceptual representations. The transfer module has the goal of transforming the representation of the input text in the source language into an equivalent representation that serves as a point of departure for a translation into the target language. A special module is necessary for each language pair as shown in Figure 4.15.

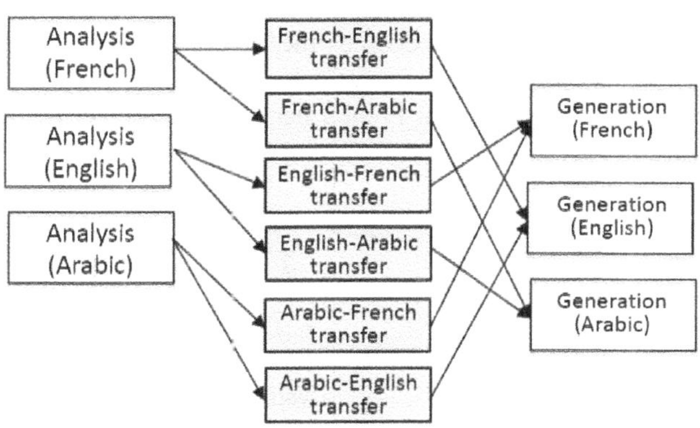

Figure 4.15. *Architecture of a transfer-based system with three languages*

Many translation systems have adopted this easily implemented approach. The complexity of the system increases considerably with an increase in the number of languages. For example, for *n* languages, we need *n* analysis modules and *n* generation modules as well as n * $(n-1)$ transfer modules. This adds up to a total of n * $(n+1)$ modules in the system. For example, for a value of *n* equal to 3, there is a total of 12 modules. For *n* equals 6, the total number of modules is 42, 30 of which are transfer modules.

The syntactic transfer is a particular form of this approach. It consists of connecting two surface syntax structures, one in each language. Three main steps are necessary to realize a translation system in this case. First, the syntactic analysis tree of the input sentence is constructed. Then, the constituents are rearranged. Finally, the words are translated and the tree of the sentence is constructed in the target language. In the example in Figure 4.16, the syntactic parsing of the input group is shown. The syntactic transfer is a rearrangement of the elements in the sentence.

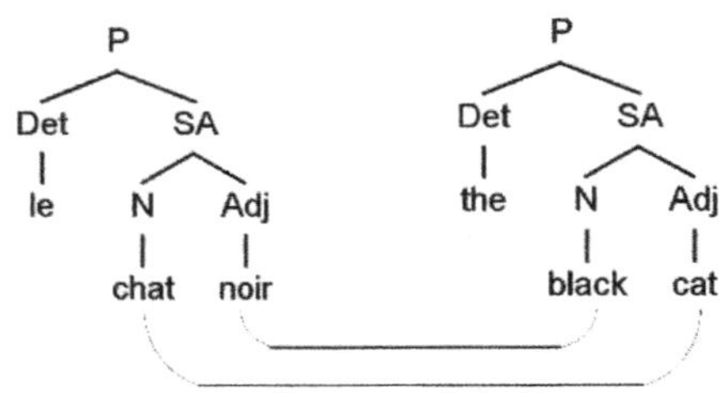

Figure 4.16. *An example of a syntactic transfer*

In reality, it is not always this simple. Often, the translation results in deeper changes. For example, the subject becomes the object, or vice versa. Consider the following pair: *You(subject) like [the house](object), [La maison](subject) te(object) plaît*. The subject of the sentence in French becomes the object of the sentence in English, as shown in Figure 4.17.

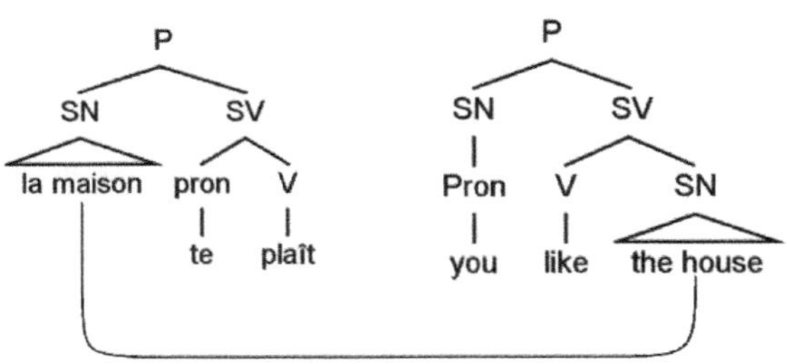

Figure 4.17. *Syntactic transfer with order inversion*

The introduction of syntactic knowledge resolves issues with the order, but the semantics of the translations obtained are often inappropriate, in particular because the formulation of fixed expressions from one language to another can pose problems for this kind of translation.

Semantic transfer approaches consider the translation as a relation between two semantic representations of the input sentence and the output sentence, respectively. Several semantic formalisms have been used such as the QLF (quasi-logical form), based on predicate logic. To make the syntax–semantics interaction more explicit within the LTAG formalism, [SHI 90] proposed parallelizing the syntactical structure, represented by elementary trees, and an argument predicate structure that serves as a semantic interpretation skeleton for the elementary tree with which it is associated. The semantic representation also has the form of a tree structure. Thus, at least one semantic tree is associated to each elementary tree and the links are defined between the nodes of the two trees that restrict the possible derivations. The main innovation of this formalism is that the syntactic and semantic derivations must be synchronized. Thus, the derivation of two trees $<\alpha_1,\alpha_2>$ follows these steps:

1) In a non-deterministic way, choose a link between two nodes (n_1 to α_1 and n_2 to α_2).

2) In a non-deterministic way, choose a pair of trees $<\beta_1,\beta_2>$ of the grammar.

3) Create the pair $<\beta_1<\alpha_1, n_1>, \beta_2<\alpha_2, n_2>>$ where $\beta<\alpha, n>$ is the result of a primitive relation on α at node n using β.

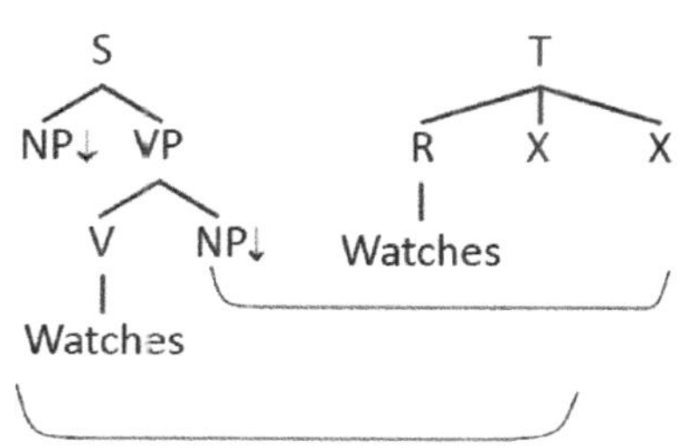

Figure 4.18. *Example of a synchronized syntactic tree and semantic tree*

Machine translation is the most common application of this formalism. The basic principle of these applications is to use the transfer rules from one language to another. For each derivation tree in the source language, a corresponding derivation tree is constructed in the target language. This is done by establishing a link between each node on both sides and by preserving the relations of dominance between the nodes in the source derivation tree. The transfer diagram is presented in Figure 4.19.

A modified version of this formalism has been proposed for the automatic translation of speech [CAV 98].

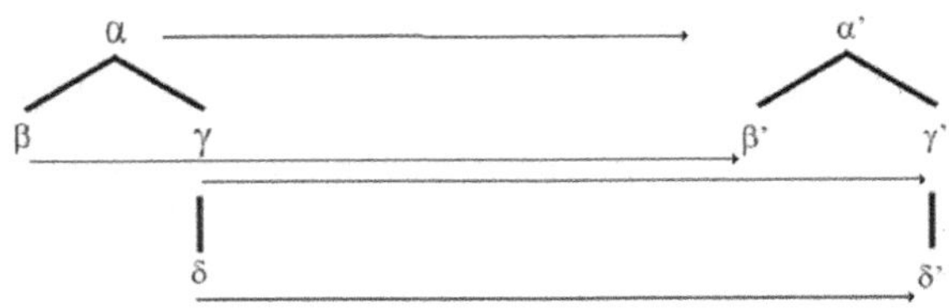

Figure 4.19. *Diagram of a simple transfer [PRI 94]*

From an engineering perspective, transfer-based approaches are not an elegant way to carry out translation. As shown, there are many modules for each language pair. Despite this design problem, this approach remains a fast way to construct translation systems. The transfer-based approach closely resembles the pivot-based approach if attempting to obtain deep linguistic representations.

4.2.5.3. *The pivot or interlingua approach*

In 1629, René Descartes proposed this approach using mathematics as a universal language. It was this very principle that was taken up in the second half of the last century as a solution to machine translation problems. In the interlingua approach, processing occurs in two steps, as shown in Figure 4.20:

– *An analysis module* that produces a representation of the input text in the source language in a pivot language (designed to be independent of all language).

– *A generation module* that generates a well-formed sequence of words from the semantic representations obtained in the pivot language. An automatic translation system involves practically all components of linguistic knowledge: morphological analysis, syntactic analysis and semantic analysis.

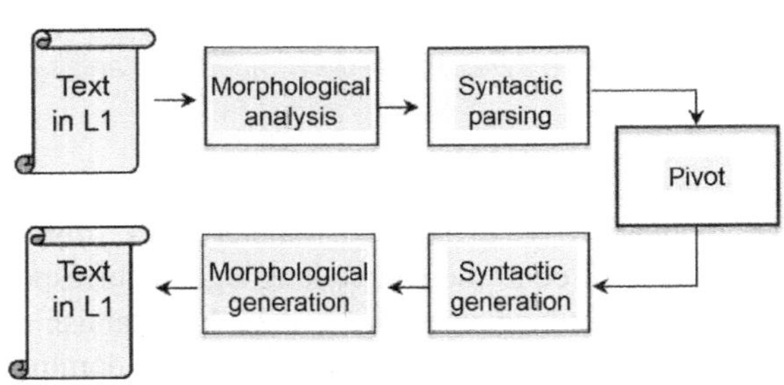

Figure 4.20. *Architecture of pivot-based systems*

Contrary to semantic transfer approaches, pivot-based analysis also involves pragmatic knowledge that allows for a better consideration of the context. However, this depth also has a cost, which is the requirement of analysis at several levels. On the one hand, this weighs down the development process considerably (requires more work). On the other, in terms of processing, this increases the translation time, which is crucial for some applications (especially in speech translation, where the translation must be instantaneous). An example of a pivot from the European project NESPOLE is given in Figure 4.21.

AGENT: a double room costs 150 a night.
Une chambre double coûte 150 dollars la nuit.
a give-information + information+room
Room type=double, price=(quantity = 150,
currency=dollar,
per-unit = night)

Figure 4.21. *Example of a pivot in the domain of the translation of hotel reservations*

Adopted in several machine translation systems (ARIANE by GETA, 1961–1971 and CSTAR, 2000s), this approach appeals to researchers for several reasons. On the one hand, it is ideal for typologically distant language pairs like (French–Arabic) and (English–Chinese). On the other hand, it stands out for its economy because the pivot can be developed once for all (regardless of the languages pairs intended). Finally, from a theoretical point of view, it relies on the idea, which is not unanimously agreed upon in linguistics, of the existence of language universals.

Imperfect, pivot translation suffers from limits that are known in the literature. It requires deep analyses that are difficult to implement such as the consideration of global textual phenomena, including inter-sentential phenomena. The multiplication of modules increases the sources of errors that can spread from one module to another. For example, a morphological analysis error causes a syntactic analysis error, which in turn leads to a semantic interpretation error. Due to the complexity of implementation, pivot translation is practically only used in closed domains like the translation of tourism texts like the CSTAR, VERBMOBIL and NESPOLE projects.

4.2.5.4. *Example-based translation (EBT)*

The use of search engines by human translators to find the translation of an unknown expression is a common practice today. For example, to find the English equivalent of the expression *fait pour servir et valoir ce que de droit*[15], an Internet search of some French sites can lead to the correct response. Sometimes, however, the search results in a variety of possible translations with different meanings. To transform this empirical process into an efficient algorithm, a parallel corpus should be used in practice. In this kind of corpus, the sentences are associated with their translations considered in their original context. These sentences are preferred to minimal pairs, that is, sentences where one element is changed at a time (see sentences [4.9]):

How much is that **big watch**?	Combien coute **cette grosse montre** ?	[4.9]
How much is **that yellow bike**?	Combien coute **ce vélo jaune** ?	

Example-based translation was first proposed by the Japanese [NAG 84]. Particularly appropriate for representing the context, this approach is well equipped to resolve ambiguities because verbal groups are particularly dependent on context by nature. It generally consists of a verb potentially followed by an adverb or a nominal or prepositional group, or sometimes both. The different forms that the arguments of the verb can take considerably affect its meaning and are sources of ambiguities, for example, the verb *ask* whose meaning changes significantly depending on the nature of its arguments. In general, most of the example-based translation approaches follow these steps:

– Matching: searching for fragments of the input sentence in the corpus.

– Alignment: identification of corresponding translations in the corpus. This step has the most variety between the EBT approaches. It can consist of the alignment of pairs of character strings or pairs of more complex forms: syntactic trees with context-independent grammar or formalisms like LFG and LTAG.

– Recombination: this consists of the assembly of the fragments into a correct sentence in the target language.

15 Issued for all legal intents and purposes.

From the point of view of performance, example-based systems give satisfactory results when the data used are well aligned and have good coverage of phenomena that are likely to be encountered in the target language, which is difficult to obtain except where it concerns limited applicative domains. That is why this approach is not used by commercial systems that prefer statistical approaches whose results are more easily controlled and are therefore better suited for their needs. However, EBT remains a promising domain and many teams of researchers are working on possible improvements.

4.2.5.5. *Statistical approach*

Like the example-based approach, this approach is based on prior experiences through a parallel corpus. The basic principle of statistical translation is the following[16]: suppose that there is a sentence in French *f* that we want to translate into Arabic to obtain sentence *a*. Depending on the translators, we can obtain different formulations that are more or less equivalent for *a*. The question is then to choose the best one of these forms. One possible response to this question is provided by probabilities. We can calculate the probabilities of possible translations pr(a|f) knowing the sentence *f* and selecting the more probable one. These probabilities are calculated starting from a parallel corpus. Of course, the quality of the translation model will directly depend on the quality of translations present in the base corpus. To infer probabilities from the base corpus, the Bayes formula is traditionally used (equation [4.10]):

$$p(a|f) = \frac{p(f|a)pr(a)}{p(f)} \qquad [4.10]$$

We can ignore the probability of the input sentence $p(f)$ because it is shared by all of the translations. Thus, finding a better translation comes down to finding the value of *a* that will maximize $p(f|a)p(a)$.

The idea of statistical MT is closely linked to the alignment of units of translation in the source language and the target language. These units can concern the sentence, the group or the word. To do this, several models have been proposed by IBM (see [KNI 99, LOP 08] for an overview). Finally, note that since 2006 statistical translation has had its own annual international workshop[17].

16 There is a great deal of diversity in the literature on the subject. Only the essentials are presented here.

17 http://www.statmt.org/wmt06

4.2.6. *Example of a translation system: Verbmobil*

The German project Verbmobil is, to my knowledge, the largest project ever realized in the world on the translation of speech in real time. As indicated by its name, this project is intended to process spoken conversations through mobile telephones [WAH 00b]. The main objective of the system constructed over the course of this project is the automatic translation of speech, but automatic dialogue summary functions have been added as well, for example, to generate a general reminder of what the two interlocutors have said during their negotiation and thereby validate the information exchanged in the dialogue.

Processing these dialogues occurs in an entirely centralized way in the main server of the system. As Figure 4.22 shows, the first locutor emits his utterance via his mobile phone. The signal emitted is then transmitted by satellite to the central server that carries out the automatic translation and produces a synthesized utterance in the target language that corresponds to the utterance received. Finally, the synthesized utterance is transmitted via the satellite to the addressee's mobile phone. Users only need a mobile phone to use this translation service in this model that is known as Software As a Service (SAS).

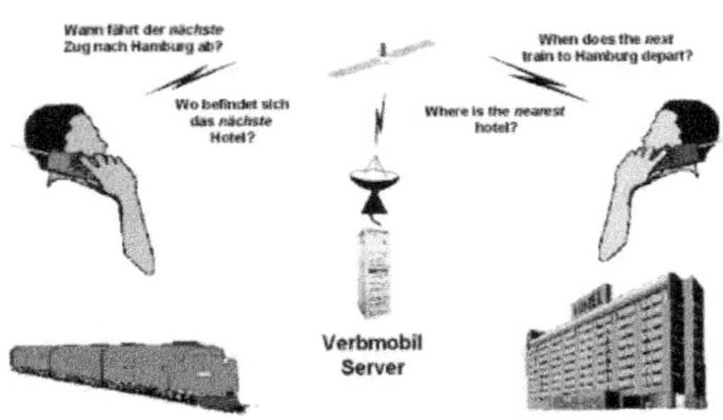

Figure 4.22. *Diagram of a mediated dialogue in Verbmobil [WAH 00a]*

One of the specificities of Verbmobil is the parallelization of different modules that have the same function and merging the results to combine their advantages without their disadvantages.

The following paragraphs will present the main components of Verbmobil that are relevant to this book. Figure 4.23 presents the system interface as well as its main functionalities, but it does not provide its real architecture, which is much more complex. The final version of Verbmobil contains 69 modules, each of which interacts with at least one other module. Given the parallel processing approach and the constraints of real-time processing, the communication requirements between the different modules are enormous from the point of view of the quantity of information. Given the non-sequential aspect of Verbmobil, this implies that the modules exchange not only the input and the output of each module but also the attempts of the high-level modules to synchronize the flow of information, constraints, alternatives, confidence scores, probabilities, etc.

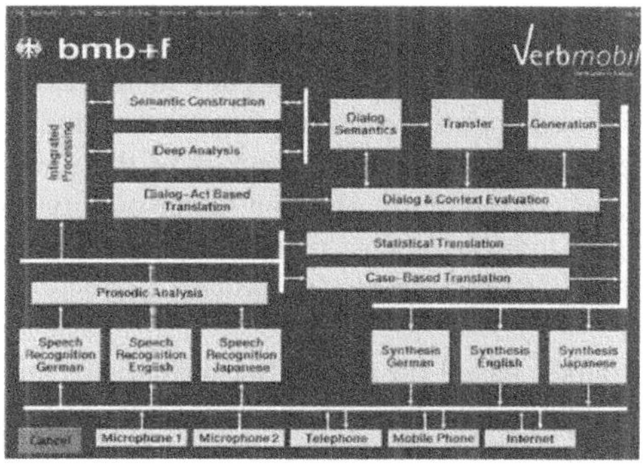

Figure 4.23. *The interface of the Verbmobil system [WAH 00a]. For a color version of this figure, see www.iste.co.uk/kurdi/language2.zip*

To meet these requirements, a multi-blackboard architecture has been implemented. This architecture contains three different types of components [KLÜ 00]. There is a set of independent modules called knowledge sources. These modules are the main element of the Verbmobil system. Unlike most blackboard-based architectures, Verbmobil uses a series of boards which are each used to represent the intermediary results of each step of the processing. Finally, there is a control module whose function is to allocate resources to optimize the runtime and module sequencing. The adoption of a central blackboard-based architecture makes it possible to add or remove modules without significant repercussions in the other modules. For example, two recognition modules for German are used to process multi-lateral conversations involving two Germans. Despite its complexity, this architecture has proven to be more effective and more appropriate than a multi-agent architecture that was used in the first phase of the project. Concerning the syntactic analysis, the system adopts a multi-engine approach to guarantee robustness. Thus, three syntactic parsers have been used in the processing: a stochastic LR parser, a chunker and a HPSG-based deep parser.

The stochastic LR parser adopts an incremental approach [RUL 00]. It was inspired, among others, by the works of Ted Brisoe at the University of Cambridge on LR parsers and their extension for processing unification-based grammars. The choice of the LR algorithm was essentially motivated by its efficiency and its adaptation for processing the word trellis in the graph provided by the recognition system. To improve the analysis quality, this module uses a post-processing phase with tree transformation rules. These rules are learned automatically from a corpus with the transformation-based learning method [BRI 93].

Based on the CASS system developed by Steven Abney, the partial parser uses memory-based learning techniques (see [DAE 05] for an introduction). In order to satisfy the constraints imposed by Verbmobil, this parser adopted an incremental approach [HIN 00]. Particular attention was paid to the assembly of segments, a subject that is addressed relatively rarely in the context of partial parsing approaches. The objective was to facilitate the task of the semantic analysis module as much as possible and consequently obtain a better-quality analysis. Compared to other approaches, the evaluation showed that this approach produced better analysis results in terms of robustness but with the shallowest analysis.

A deep syntactic parser based on the HPSG formalism, this parser was realized in the context of a collaboration between the DFKI at Saarbrücken, the CSLI at Stanford and the Language Processing Lab at the University of Tokyo [USZ 00]. One of the main problems faced by this parser was the reduction in necessary runtime. The features used to represent the linguistic constraints in the framework of the HPSG formalism require many calculations that make the parser's runtime too slow to be integrated in the framework of a real-time application. Two solutions have been combined to solve this problem: 1) elimination of features that are not central to the processing as well as disjunctive features that increase the complexity of the processing, and 2) the testing of several types of implementation like, among others, the grammar approximation of finite-state automata. Finally, this problem was overcome by researchers in the Computational Linguistics department at the University of Saarbrücken, who combined several solutions proposed by different participants and managed to satisfy the analysis time constraints, which are about 0.45 s for a 10-word utterance. As expected, the evaluation results showed that this system, compared to the two others, provides the deepest analysis but the least robust analysis of problems of recognition or extra-grammaticalities of speech (see [MÜL 00, FLI 00, SIE 00] for the final results of this parser on German, English and Japanese, respectively).

The three parsers process the same graph of words enriched with prosodic annotations produced by the recognition system. The three parsers are also guided by an A* algorithm to choose the most likely paths from a speech recognition perspective [ULR 00]. The three parsers use a specific analysis module to construct a semantic representation corresponding to the output of each one. This applies to complete parses as well as partial ones. The data exchanged between the different modules are represented in the Verbmobil Interface Terms (VIT) format, which is, as the name indicates, a special format intended to standardize the output of the parsers. The conversion of the output of the parsers into a standardized semantic representation makes if possible to carry out the post-processing operations intended to select the best analysis from the candidates [RUP 00].

As with the ALPAC report, the difficulty of the semantic approach has been known since the 1960s. Thirty years later, the statistical method has tentatively given the impression of solving the problem; however, after the leap forward that it has allowed, specialists in machine translation are now faced with the same difficulty, because although enormous progress has been made in a relatively short amount of time, the expectations of users have also progressed with the technology.

4.3. Information retrieval (IR)

4.3.1. *IR and related domains*

We are constantly making more or less important decisions on a daily basis. Of course, our decisions are generally based on information that we collect in our near or distant environment. In the case of so-called intelligent artificial systems, this happens in the same way. Every serious decision requires access to information. This information can be prepared in advance at the level of the format or the level of the content, as it can be raw, without any formatting or any constraints on the content. Depending on the nature of the information provided to the system, there are two types of systems: Information Systems (IS) and Information Retrieval Systems (IR) or search engines. The data flow diagrams (DFDs) of these systems are provided in Figure 4.24.

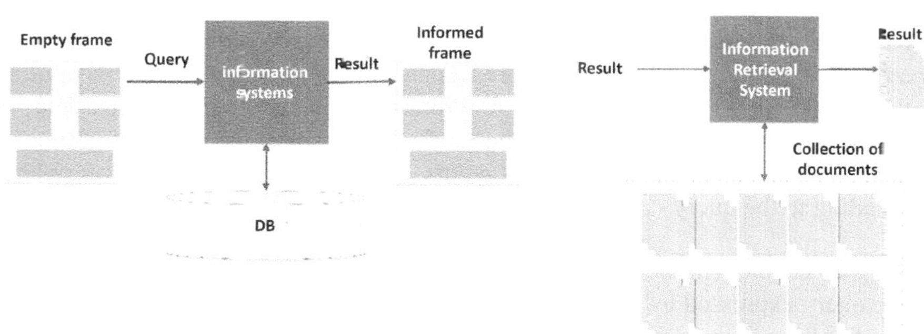

Figure 4.24. *DFDs of information systems and information retrieval systems*

Information Systems, which are increasingly common in our daily lives, are equipped with structured information frames based on which they provide the required information. The most common examples of this type are administrative management systems such as Enterprise Resource Planning (ERP) systems and Geographic Information Systems (GIS) like Google and Bing Maps that make it possible to view enriched maps from satellite images. The main task of these systems is to find a precise and structured answer to a precise and structured query with potential modifications to it. For example, a simple query for an information system consists of finding the list of customers who bought a given product, indicating for each one the quantity purchased as well as the purchase total. A modification of the previous query would consist of making a list of sales of the same product, by geographical area or customer age group. Finding such information requires storing information in tables, each of which is specifically dedicated to storing a part of the information, such as the names and addresses of customers. Interconnections between these tables make it possible to assemble the parts of the information in order to find or calculate the information requested. In terms of content, constraints can be imposed on the fields of tables like the type of values stored (date, character strings, etc.) with the universal constraint that the values in these fields all be recognizable by deterministic finite-state automata.

With the increase in power of electronic data processing, the Internet and the move to e-only document management, Information Systems are led increasingly often to also store information in natural language, images or sound signals, which are called semi-structured information. For example, this includes the answer to an open question in an inquiry, a hospitalization report written by a doctor, a photo, a video or song recording, etc. However, the analysis of these fields can no longer occur with simple deterministic finite-state automata; it requires tools that are specific to the type of information stored, such as NLP software, image recognition software and sound pattern software. With these richer and more complex data, we lose the property of structured data that is the assurance in all cases of getting an answer if the information system contains at least one field with information responding to the query.

Search engines, in turn, have the mission of finding documents that are relevant to the query expressed by a user. They can therefore be considered to be a particular case of information systems. Their specificity in relation to information systems resides in the absence of constraints on the queries, data collections and the system's output. In the case of a textual search, users can interrogate the system using a string of words, either a complete sentence or a combination of keywords that may contain meta-symbols based on a rudimentary syntax accepted by the search engine. The same applies for the documents that form the search space as well as for the output of the system, which consists of a subset of the documents in the collection. Another characteristic of search engines is the type of document collection that they include; they can be limited, as in the case of a documentary information retrieval system, or completely open, as in the case of the Internet.

There are two major functionalities in a search engine. Indexing is a step that occurs offline and consists of constructing a simplified representation of the collection in order to reduce the space necessary for its storage while making access to information faster. Indexing is done on both the basis of the content of each document and its interconnections through hyperlinks to other documents. The latter information is used to calculate the page rank, which is a global score of the importance of a document that makes it possible to distinguish two documents that are related in a similar way by the hyperlinks to other documents in the collection. More generally, the page rank is a measure of the centrality of a node in a graph.

The other major functionality of a search engine is the search in a collection for documents that are relevant to the query and the presentation of results. This is generally done in the form of a list arranged in order of decreasing relevance. The response documents are generally presented as hyperlinks associated with a small extract of the document or a summary. Other forms are also possible, such as a navigable graph, where each node represents a document, a topological map or lists of icons, generally used for image search engines.

For a general introduction to information retrieval, see [AMI 13, KOW 11, FRA 92, MAN 08].

4.3.2. *Lexical information and IR*

All texts, regardless of type, are formed by a sequence of words that are aligned according to constraints that are morphological, syntactic, discursive, etc. It is therefore natural to start by understanding the contribution of lexical knowledge to information conveyed by the text and, consequently, to possible simplifications that can occur at this level.

4.3.2.1. *Zipf's law*

In the domain of information retrieval, we typically search for texts that are most relevant to a query formulated as one or more keywords. This comes down to searching for texts where these words have a significant role. The occurrence of a word in a text is not enough to judge that this word has a privileged role in this text. To understand the measure of importance of a word in a text, Zipf's law and its derivatives are essential.

In his quest to formalize the principle of minimal effort that is supposed to explain human behavior, Zipf proposed an empirical law called Zipf's law [ZIP 49]. This law stipulates that, in a given corpus, the frequency of a word is inversely proportional to its rank in a frequency table. In other words, the most frequent word in the corpus appears two times more than the next word in the frequency table and three times more than the third word and so on. This observation can be noted by equation [4.11]:

$$f \propto \frac{1}{r} \quad [4.11]$$

From equation [4.11], it is clear that for every corpus, there is a constant k such as $k = \frac{1}{r}$. The function of mass in Zipf's law is given by equation [4.12], where the symbol N corresponds to the number of words, the symbol k corresponds to the rank and s > 0 corresponds to the value of the exponent that characterizes the distribution:

$$f(k, S, N) = \frac{1/k^{s}}{\sum_{n=1}^{N}(1/n^{s})} \quad [4.12]$$

To verify Zipf's law empirically, consider a collected set of French novels from different genres and different periods that I call Corpus of Selected French Novels (CSFN). This collection contains about 1.5 million words from the following 10 literary works: *A la recherche du temps perdu* by Proust, *La comédie humaine* by Balzac (three volumes), *La princesse de Clèves* by Madame de la Fayette, *Germinal* by Émile Zola, *Voyage au centre de la terre* by Jules Verne, *Le rouge et le noir* by Stendhal, *Madame Bovary* by Gustave Flaubert and *Notre-Dame de Paris* by Victor Hugo. Table 4.2 contains the 50 most common words, their frequencies (f), their ranks (r) and the value f*r that is supposed to be fixed throughout the corpus. As we can see, the values of (f*r) are not fixed as predicted by Zipf's law. However, a tendency toward the stabilization of these values starts to be seen as of rank 15 with *k* close to 250,000. Note that [MAN 99] observed similar phenomena for the novel Tom Sawyer in English, which means that these observations are not unique to CSFN corpus or the French language.

Thus, it can be said that Zipf's law corresponds to a general tendency rather than a real law that can be verified in all cases. Drawing the curve of function of the equation is a good way to observe the Zipf *effect* on a larger scale. The curve of the function for the 300 most frequent words in the CSFN corpus is given in Figure 4.25. This curve shows that there is a small group of words whose frequency is very great, and at the same time, there is a large number of words whose frequency is not very high. This can be clearly seen by the rapid merging of the curve with the x-axis.

Word	Frequency	Rank	(f*r)	Word	Frequency	Rank	(f*r)
de (of)	70,982	1	70,982	ce (this)	9,738	26	253,188
la (the; fem. sing)	42,268	2	84,536	s' (refl. pronoun)	9,695	27	261,765
et (and)	32,821	3	98,463	est (is)	9,336	28	261,408
le (the; masc. sing)	32,296	4	129,184	son (his or her)	9,240	29	267,960
à (to)	30,435	5	152,175	n' (not)	9,045	30	271,350
il (he)	26,459	6	158,754	était (was)	8,975	31	278,225
l' (the; sing)	25,565	7	178,955	au (at the)	8,721	32	279,072
les (the; plur)	25,118	8	200,944	vous (you; plur)	8,586	33	283,338
un (a; masc)	22,043	9	198,387	sa (his or her)	8,479	34	288,286
d' (of)	20,839	10	208,390	avait (had)	8,410	35	294,350
en (in)	18,760	11	206,360	plus (more)	8,305	36	298,980
que (that)	18,684	12	224,208	par (by)	8,079	37	298,923
une (a; fem)	17,637	13	229,281	sur (on)	7,190	38	273,220
elle (she)	16,194	14	226,716	avec (with)	6,369	39	248,391
des (some)	15,217	15	228,255	comme (like)	6,354	40	254,160
qui (who)	15,214	16	243,424	a (has)	6,230	41	255,430
qu' (that)	13,471	17	229,007	mais (but)	6,174	42	259,308
ne (not)	12,919	18	232,542	si (if)	6,150	43	264,450
dans (in)	12,617	19	239,723	ses (his, her or their)	6,055	44	266,420
je (I)	12,096	20	241,920	cette (this)	5,945	45	267,525
se (refl. pronoun)	11,715	21	246,015	on (we)	5,738	46	263,948
pas (not)	11,360	22	249,920	tout (all)	5,215	47	245,105
du (of the)	10,647	23	244,881	y (there)	5,083	48	243,984
pour (for)	10,180	24	244,320	dit (said)	5,040	49	246,960
lui (him or her)	9,797	25	244,925	nous (we)	4,627	50	231,350

Table 4.2. *List of the most frequent words in the CSFN*

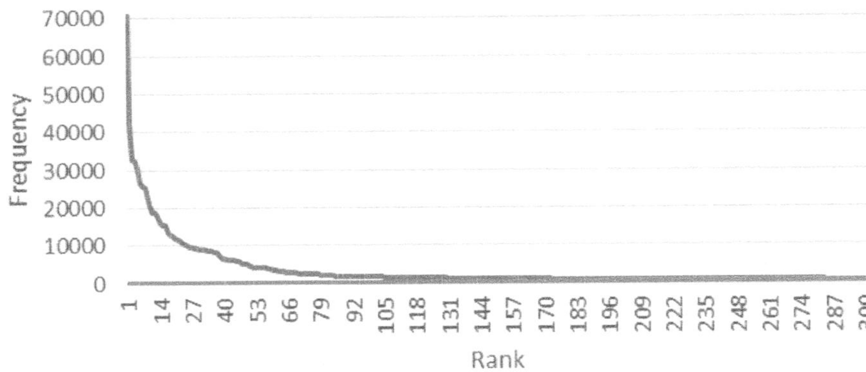

Figure 4.25. *Curve of the relation between frequency and rank for the first 300 words in the CSFN corpus. For a color version of this figure, see www.iste.co.uk/kurdi/language2.zip*

It is natural that such a strong declaration about linguistic systems attracted so much interest from among researchers, especially in the domains of applied mathematics. Zipf's law has been refined many times, the most well-known of which is probably Mandelbrot's law [MAN 57].

4.3.2.2. *Filtering empty words*

As can be observed in Table 4.2, the most frequent words are grammatical: prepositions, determinants, pronouns, particles, etc. Despite their linguistic value, these words are so common in all documents that they have practically no distinctive role. To reduce the size of the documents and therefore the calculation cost and the space necessary to store the data, some researchers recommend simply removing these kinds of words, which are commonly called empty words, stop words or anti-lexical words. The lack of theoretical foundation makes the creation of such lists an essentially empirical and ad hoc process. This explains the presence of a multitude of lists of this kind with fundamental differences, as much from a quantitative point of view as from a function point of view (categories of words concerned). However, it must be admitted that grammatical words are an essential tool for search engines, for example, to detect keyword stuffing that occurs by repeating these words in a text to bias its indexing.

4.3.2.3. *Weighting words*

The simplest method of indexing consists of associating each document in a collection with a list of keywords that are included in it. That way, the search engine simply says that a document does or does not contain the keyword in question. This is called Boolean indexing. This view is said to be naive because it does not consider the importance of the role of each keyword in the document. This role is particularly important when there are a large number of documents. Moreover, it presupposes the existence of a large list of keywords whose creation requires considerable efforts especially for large documents where the content is likely to change. To reflect the importance of a word in the text, the simplest way would be to consider the Term Frequency (tf). In other words, the more frequent a word is in a text, the more important is its weight. Again, this is not sufficient because some words, not necessarily grammatical words, are more frequently used in some registers than others, which means that the discriminating capacity of these words is not very great.

One way to consider the importance of a word at the document level while considering its frequency on a global level would be to examine what is called the Collection Frequency (cf), which is the frequency of a word throughout the entire corpus. Before going further, consider Table 4.3, which includes a comparison between the frequencies of certain terms in the novel *Germinal* by Emile Zola in opposition to their frequency in the CSFN corpus, which includes 10 novels.

Word	Germinal (tf)	CSFN (cf)	tf/cf*100	idf	tf-idf
Froid (cold)	62	330	18.7%	10	0
Mine (mine)	57	145	39.3%	10	0
Houille (coal)	50	54	92.5%	3	26.14
Herscheuse (female worker who pushes a cart)	24	24	100%	1	24

Table 4.3. *Comparison tf, cf, tf/cf, idf and tf-idf scores*

The word *cold* is a relatively frequently used word in *Germinal* (rank = 321). It also frequently occurs in other documents in the collection and therefore has a limited discriminatory power of only tf/cf=0.187. As it has a generic meaning, this word does not belong to a specific semantic field and does not play a particular semantic role at the global level of the corpus. Because the events of *Germinal* take place in the world of mining, it would at first seem tempting to think that the word *mine* itself would be characteristic of this novel. The results show that this word is frequent at the local level and the collection level. This frequency at the collection level can be explained by the highly polysemic character of this word: in the Petit Robert (French–French dictionary), there are eight meanings arranged in two different entries. On the other hand, the word *houille* (coal), which is relatively frequent in *Germinal*, is not frequent at the collection level, which gives it an important distinctive role. Finally, although the word *herscheuse* (female worker who pushes a coal cart) is not very frequent in *Germinal* (rank = 751), it is very characteristic of this novel because it does not appear in any other novel in the collection.

To push this weighting method even further, another important factor should be considered, which is the size of the collection as well as the number of documents in which the term appears. The high frequency of a term in a collection of a few documents does not have the same significance as in a collection with several thousand documents. Similarly, the collection frequency does not give an idea of the distribution of the term throughout the collection: the word might appear in only a few documents or in all of them. To consider this information, some researchers consider the Inverse Document Frequency (idf) multiplied by the local frequency of the term *tf* [ROB 76, ROB 04]. The global score of a term *t* in a document *d* is therefore calculated using equation [4.13], which corresponds to what is called the Term Frequency-Inverse Document Frequency (tf-idf), where *N* is the number of documents in the collection and *fdt* is the number of documents in the collection where the term *t* appears:

$$s(t,d) = tf \times \text{idf} = tf \times log\frac{N}{fdt} \qquad [4.13]$$

Returning to the CSFN corpus, the value of *N* is equal to 10. For example, the word *froid* (cold) appears in all documents in the collection, so fdt =10. The score of this word in *Germinal* is obtained using the equation: $S(froid, Germinal) = 62 \times \log\frac{10}{10} = 0$. As noted, the tf-idf method considers that words that appear in all of the documents in the collection do not have a distinctive power and attribute them zero weight, independently of their frequency. Similarly, this method is very generous with words that only appear in a limited number of documents even if their frequency is not very great: *herscheuse* has a similar weight to *houille,* even though *houille* has a *tf* that is almost two times greater.

Many kinds of frameworks have been proposed to further optimize word weighting. For a review of some of these frameworks, see [SAL 88, LAN 05, CUM 07, NAN 04].

4.3.2.4. *Lemmatization*

Lemmatization is another way to simplify a document. It consists of grouping all of the inflectional variants of a word into a single item that serves as its identifier in a dictionary. Thus, the singular and plural forms of a noun will be reduced to its singular form or all conjugated forms of verb in all tenses and all modes will be associated with the infinitive form. The advantage of lemmatization in the context of information retrieval is that it reduces the lexical forms considered for the search.

4.3.3. *Information retrieval approaches*

4.3.3.1. *Scanning the complete file*

The simplest method to establish a link between a query and a document is to search for documents that contain this word in the entire database of available documents. This is made easy by using one of the efficient search algorithms proposed, for example, by [BOY 77, SUN 92]. The major disadvantage of this method is that it provides a very superficial analysis of the query–document relation. For example, there is no distinction between words that play an important role in distinguishing between two documents and words whose distinctive role is limited. Moreover, searching through all of the documents significantly limits the processing speed. This kind of solution was used at the end of the 1990s, particularly in the filters of classic messaging systems like Pegasus Mail and Outlook Express.

4.3.3.2. *Inverted lists*

According to this method, each document can be represented by a series of keywords. In order to accelerate the search process, the inverted file is built. This provides a list of keywords from all of the documents, ordered alphabetically or according to their global frequencies, and each associated with the documents in which it appears. Typically, grammatical words are removed from this list to reduce its size and make the search more efficient. To increase the efficiency of this approach even more, double-pass solutions have been proposed. In the second level, all of the words starting with the first two letters are stored in the same place, while the first level contains hyperlinks to the second level. The advantages of this approach are its easy implementation, fast searching capacity and the possibility of easily processing synonyms. The disadvantages include the size of the inverted file, which can sometimes be quite large, and the difficulties related to updating modifications to documents, especially in the case of a dynamic environment.

4.3.3.3. *Signature files*

This approach consists of reducing the document to a sub-document containing only information that is relevant to the search. The document that results from the filtering process is stored in a separate file called the signature file, which is much smaller than the original document and consequently can be searched more quickly. Like the other approaches in information retrieval, it starts by simplifying the document. Grammatical words are eliminated and all uncommon words are reduced to their lemmas. Next is the file segmentation phase. Several techniques are possible for this operation, including digital techniques and n-grams. The idea of this segmentation is to identify the relevant words by segments and not at the global level of the text in order not to fall into generalities. Other approaches have gone further by proposing signature file processing on two levels. In the first level, the files are represented in the form of trees. In the second level, the information within the files are divided according to their frequencies.

The main problem with this method is finding an ideal compromise between the detail and the speed of the search: the larger the size of the signature file, the more accurately it represents the original document, but the less it allows for fast searching. As noted in [CHR 84], signature files require between 10% and 15% of the size of the initial file, whereas inverted files require between 50% and 300%. It is also much easier to add new information to the document with signature files than with inverted lists, which potentially require a new rearrangement of information for each addition.

4.3.3.4. *Clustering-based approaches*

Approaches based on clustering are the most common form of unsupervised learning algorithms. Unlike supervised methods, clustering does not require the intervention of a human expert to label the data first to guide the learning algorithm.

The basic idea of clustering in the context of information retrieval is that similar documents are assembled to form a group of the same type. The main motivation for this choice is that, generally, the documents forming a cluster tend to share the same queries or respond to the same needs in terms of information. The cluster can be used for documents as well as words. In the second case, the cluster can be used for the automatic construction of a thesaurus. For more information about clustering, see [AND 07, ROS 06, STE 00].

There are two types of clustering: flat clustering and hierarchical clustering. In flat clustering, the clusters obtained have equal importance, whereas in hierarchical clustering, the clusters hold dependency relations in the form of trees. Figure 4.26 depicts the operation of a flat clustering of a set of documents, which shows three classes of documents.

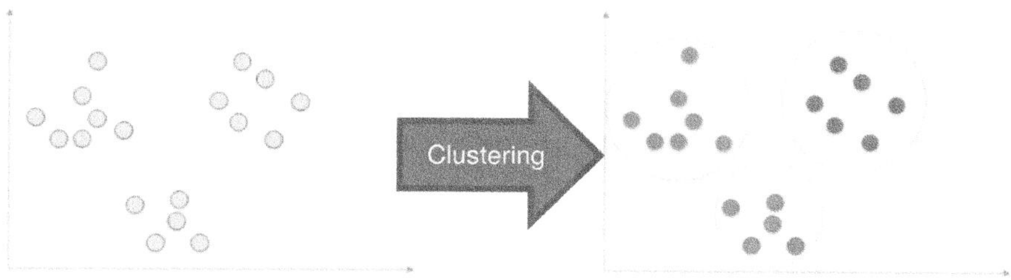

Figure 4.26. *Flat clustering of a set of documents. For a color version of this figure, see www.iste.co.uk/kurdi/language2.zip*

Clustering approaches are based on the transformation of texts analyzed in the form of a point vector. First, a series of preparatory processing steps are carried out, like the removal of stop words, lemmatization and clustering words in conceptual classes using a thesaurus. This pre-processed document is represented by a t-dimension vector, where t is the size of the indexing vocabulary. The vectors used can be binary or not. The absence of an indexing term is marked by a 0 and its presence is marked by a 1 or by a positive number indicating the importance (weight) of the concept in the classification of the document. Different functions have been proposed to determine the weight of a term:

– $FREQ_{ik}$: this function considers the frequency of a word k in a document i. It is easy to obtain and allows for greater results than binary vectors.

– The specificity of a term: logN – $\log(DOCFREQ_k)+1$, where $DOCFREQ_k$ is the number of documents that contain the term k and N is the total number of documents. This function is relatively easy to obtain and more powerful than binary vectors.

– The inverted frequency of a document: $FREQ_{ik}/DOCFREQ_k$; experimental results have shown that this function is more effective than the previous one.

– $FREQ_{ik} * TERMREL_k$, where $TERMREL_k = r_k/(R - r_k)\ s_k/(I - s_k)$. The term $TERMREL_k$ represents the relevance factor, r is the total number of relevant documents that contain the term K, I is the total number of non-relevant documents and s_k is the number of non-relevant documents that contain the term k. Experimental studies have shown that this method is the best [SAL 83].

After having represented the documents in a t-dimensional vector, the next step consists of dividing the points obtained into clusters. This operation must satisfy the following set of diverse criteria:

– The method must be stable as the documents increase in number. In other words, the distribution of points should not radically change with the insertion of new documents.

– Small modifications in the description of the documents should lead to small modifications in the distribution.

– The method should be independent of the initial order of documents.

– The algorithmic efficiency of the proposed methods, particularly the speed of the algorithm.

Several types of methods have been proposed in the literature to satisfy these constraints, including similarity-based methods, iterative methods and Bayesian methods, which will be discussed in the following sections.

Finally, searching in a set of documents analyzed in clusters is simpler than generating the clusters. The operation consists of representing the query as a t-dimensional vector and comparing it to the centroids of the cluster. The similarity between the cluster and the query is detected using a special function: the cosine is the most commonly used function (see section 4.3.3.8 for an example).

4.3.3.5. *Similarity-based approaches*

These approaches use techniques based on graphs and require a $O(n^2)$ time or sometimes more (where n is the number of documents). The basic principle behind this technique is to calculate the similarity (distance) between two documents. The first step consists of choosing a similarity function between two documents that makes it possible to measure the distance between two documents represented in the form of a matrix. In the literature, a series of similarity functions have been proposed and give approximately equivalent results if the documents are properly standardized [SAL 83].

Suppose that we have a set of documents $D_1, \ldots. D_m$ and a set of terms that are employed in these $T_1, \ldots. T_n$. For each term T_i and document D_j, we have a weight W_{ij} that indicates the weight of this term in the document. An example of a similarity function between two documents would be the sum of the multiplication of the weight of the terms in these two documents (see equation [4.14]):

$$\mathrm{sim}(\mathrm{D_i}, \mathrm{D_j}) = \sum_{t=1}^{n} W_{it} * W_{jt} \qquad [4.14]$$

Consider the corpus of six micro-texts presented in Figure 4.27.

Doc_{01} Tourists adore visiting the Bastille in Grenoble.
Doc_{02} John adores visiting the Louvre in Paris.
Doc_{03} Sports are very good for the health.
Doc_{04} Sports classes are obligatory at school.
Doc_{05} Tourists must have health insurance before getting their tourist visa.
Doc_{06} Road accidents are the leading cause of death in our country.

Figure 4.27. *Collection of documents*

To determine whether two documents are similar or not in a corpus like the one in Figure 4.27, first we conduct two pre-processing operations: filtering stop words and lemmatization. For the sake of simplicity, we will skip the step of conceptual clustering with a thesaurus. The result is the documents in Figure 4.28.

Doc_{01} tourist, ador, visit, Bastille, Grenoble.
Doc_{02} John, ador, visit, Louvre, Paris.
Doc_{03} sport, good, health.
Doc_{04} class, sport, oblig, school.
Doc_{05} tourist, insurance, health, visa.
Doc_{06} accident, road, cause, death, country.

Figure 4.28. *List of keywords retained in the documents in this collection*

Next, we create a term-document matrix where we consider the frequency of a word in a document as an indication of its weight. For the scope of this presentation, we will draw up a partial term-document matrix for 10 terms.

To calculate the symmetry between two documents, according to equation [4.14], we multiply the weight of the terms in the documents. Because the weight of terms that are not shared is equal to zero, calculate the sum of the multiplications of the weights of the shared terms. Consider the following examples of similarities:

sim(Doc01, Doc02) = (1*0)−(1*1)+(1*1)+.... = 2.

sim(Doc01, Doc05) = (1*2)−(1*0) + (1*1)+...= 2.

sim(Doc01, Doc06) = 0.

	Tourist	**ador**	**visit**	**Bastille**	**Grenoble**	**John**	**Louvre**	**Paris**	**sport**	**good**
Doc$_{01}$	1	1	1	1	1	0	0	0	0	0
Doc$_{02}$	0	1	1	0	0	1	1	1	0	0
Doc$_{03}$	0	0	0	0	0	0	0	0	1	1
Doc$_{04}$	0	0	0	0	0	0	0	0	1	0
Doc$_{05}$	2	0	0	0	0	0	0	0	0	0
Doc$_{06}$	0	0	0	0	0	0	0	0	0	0

Table 4.4. *Term-document matrix*

By calculating all of the similarities, we obtain the document–document matrix in Table 4.5. It consists of a symmetrical matrix whose diagonal is empty because it corresponds to the relation of the document with itself.

	Doc$_{01}$	**Doc$_{02}$**	**Doc$_{03}$**	**Doc$_{04}$**	**Doc$_{05}$**	**Doc$_{06}$**
Doc$_{01}$		2	0	0	2	0
Doc$_{02}$	2		0	0	0	0
Doc$_{03}$	0	0		1	0	0
Doc$_{04}$	0	0	1		0	0
Doc$_{05}$	2	0	0	0		0
Doc$_{06}$	0	0	0	0	0	

Table 4.5. *Document–document matrix for the collection in Figure 4.27*

After calculating the similarity matrix, a similarity threshold is chosen starting from which two documents are considered to be linked. In this example, given the limited size of the documents, it has been decided to consider that two documents are similar if their weight of resemblance is at least equal to one (see Table 4.6).

	Doc01	Doc02	Doc03	Doc04	Doc05	Doc06
Doc01		1	0	0	1	0
Doc02	1		0	0	0	0
Doc03	0	0		1	0	0
Doc04	0	0	1		0	0
Doc05	1	0	0	0		0
Doc06	0	0	0	0	0	

Table 4.6. *Standardized document–document matrix (binary)*

The connected graph representing the relations between the texts is the final product of this step (see Figure 4.29).

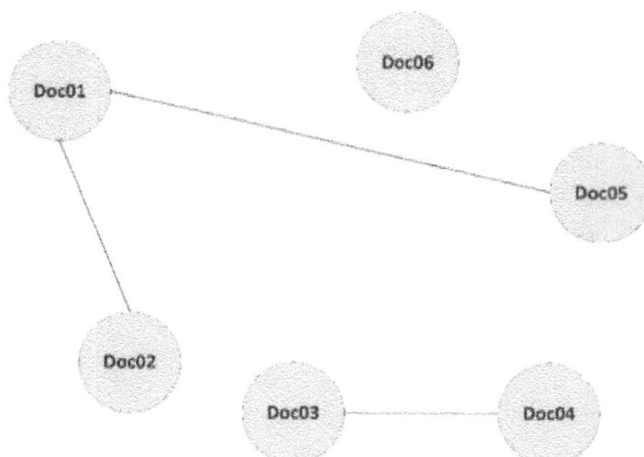

Figure 4.29. *Graph of connections between the documents*

The next step consists of applying a clustering algorithm like the connected components algorithm. This algorithm requires a similarity between the terms of a cluster. In a graph, a connected component is a set of nodes that are accessible to each other. The steps of this algorithm are given in Figure 4.30.

Computationally advantageous, this algorithm produces classes whose members are weakly classified. These classes are not strongly equivalent to concepts. Regarding the performance, this algorithm tends to be good for precision but not as good for recall.

1) Select a document that is not yet classified and put it in a class. If there are no more non-classified documents then stop the classification.

2) Put all other similar documents in this class.

3) Repeat step 2 for all documents.

4) When there are no more documents to process with step 2, return to step 1.

Figure 4.30. *Connected components algorithm*

Note that the Class_{01} includes documents that do not share any terms between them, like doc_{01} and doc_{03}.

The result of the classification obviously depends strongly on the algorithm used. It depends on the similarity function between documents and the way in which it is calculated. For example, the Stars algorithm takes an initial document that is called the kernel, considered to be the node of nucleus of the class, and then searches for documents that are similar to the seed [ASL 82]. If no more documents similar to the current kernel are found, other kernels are searched and so on. Note that the classes obtained can vary according to the initialization of this algorithm. If the documents are arranged in ascending order, it provides the following classes:

$\text{Class}_{01}(\text{doc}_{01}, \text{doc}_{02}, \text{doc}_{05})$

$\text{Class}_{02}(\text{doc}_{03}, \text{doc}_{04})$

$\text{Class}_{03}(\text{doc}_{06})$

Classes 01, 02 and 03 do not share any documents. This is a hard cluster. When two or more clusters share one or more documents, it is called a soft cluster. The main disadvantage of this method is that it requires at least one constant found empirically, which is the reconciliation threshold of texts and sometimes it is also necessary to specify the number of clusters *a priori*. These constants impose structural constraints on the data, and consequently they affect a search engine's performance considerably.

4.3.3.6. *Iterative approaches*

These methods are based directly on the description of the text, based on a hashing key, without calculating an inter-document similarity. The algorithmic complexity of this approach is very good, in general $O(n\log n)$ or $O(n^2/\log n)$. This efficiency is obtained to the detriment of certain constraints like sensitivity to the order in which the texts were analyzed or the impossibility of predicting the error rate in the description of the document. The general functioning of this approach follows these two steps:

– determination of the initial distribution;

– reiteration and association of the document to clusters, until there are no more good associations to make.

Finally, it should be noted that hybrid methods can be used. The iterative method can be used to partition documents into clusters and then the graph method can be used to divide each of the clusters obtained.

4.3.3.7. *Bayesian approaches*

These approaches are based on applying a Bayesian network to a classification task. The network contains a node *C* for the class variable and a node X_i for each feature describing an individual to order. Given a specific case *X*, the Bayesian network makes it possible to calculate the probability: $P(C = c_k \mid X = x)$ for each possible class c_k. The optimization of the classification occurs by selecting the class c_k for which the probability is maximized. Although all forms of Bayesian networks can be used for the classification, empirical results have shown that networks where all of the variables are directly connected give better results than those in "naive" Bayesian networks, where the variables are completely separate (see Figures 4.31).

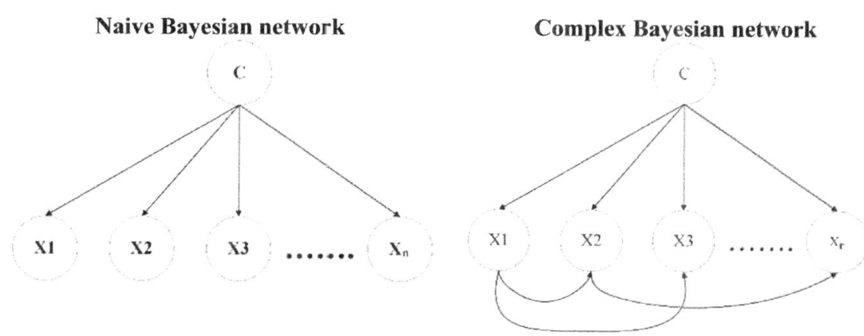

Figure 4.31. *Diagrams of naive and complex Bayesian networks*

In order to extend the capacities of naive Bayesian networks, represented by the classic equation: $P(X|C) = \Pi_i P(X_i \mid C)$, several solutions have been proposed, including the TAN algorithm by Friedman and Goldszmidt [FRI 96] and the KDB algorithm by Sahami [SAH 96]. The TAN algorithm requires that each node has a maximum of one additional parent, and in this case, the algorithm can find the optimal solution in quadratic time. Sahami's KDB algorithm in turn is based on a heuristic search of the best structure, which can potentially be sub-optimal. It can be used to find classifiers, where each node has a maximum of k parents, for a random value of k. Essentially, this algorithm chooses the k other features as parents of a node X_i, of which X_i is the most dependent. This approach has been used by Sahami for the classification of unwanted mail, or spam, in the context of the SONIA project and seems to give satisfying results for this task [SAH 98a, SAH 98b].

Aside from the approaches just presented, there are also the self-organizing maps developed by Kohonen, which is a generic neural network technique that can be applied to a cluster of documents [KOH 97], as well as the clique algorithm [AGR 98] and k-means [LLO 82, FOR 65].

4.3.3.8. *Vector approaches: latent semantic analysis (LSA)*

Vector approaches consist of constructing a representation of the texts in a collection as well as the query in the form of vectors to be able to calculate the distances between documents, on the one hand, and those between the queries and the documents, on the other hand. One of the most common vector approaches is latent semantic analysis (LSA).

LSA is both a theoretical approach and a method for extracting and representing information relative to word usage from a large corpus [LAN 97, LAN 98, BER 95]. It was initially developed to solve the problem of synonymy in the domain of information retrieval. According to this approach, the meaning of a text can be represented by a matrix that considers the occurrences of all the words in that text as well as the (co-)occurrences of these words in other possible contexts in other texts. As a cognitive model of human intelligence, several works have shown that LSA parsers make it possible to simulate the acquisition and recognition of vocabulary [FOL 96, ALB 08]. Finally, from an applicative point of view, as will be presented later in this text, latent semantic analysis and its variant latent semantic indexing (LSI) have been used notably in the domain of indexing and information retrieval, but there are also applications in other domains, such as intelligent tutoring systems [GRA 05, MIL 03, KU 11].

The processing principle is simple. It calculates and stores the reduced vectorial representation of all of the documents of the collection offline. When a user submits a query, it calculates the vectorial representation of this query in the same way and then it displays the documents whose vectorial representation is the closest to that one, in ascending order of distance (see Figure 4.32).

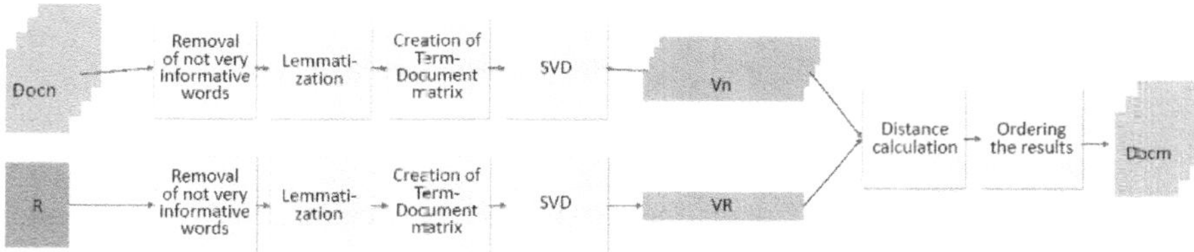

Figure 4.32. *General diagram of information retrieval with LSA. For a color version of this figure, see www.iste.co.uk/kurdi/language2.zip*

One of the main objectives of LSA is to reduce redundancies in the documents as much as possible in order to facilitate the search process and reduce the space necessary for storing information. To do this, the processing starts by removing words that are determined to have little informative value or empty words. Then, the retained words are lemmatized and used to construct term–document matrices. In this matrix, each entry consists of an integer representing the number of occurrences of a specific term in a given document. The singular value decomposition (SVD) of the matrix obtained in the previous step is then calculated. Consider the term–document matrix X with t lines (one for each term that appears in the chosen document) and d columns (one for each document). The result of the SVD is calculated according to equation [4.15]:

$$X = T_0 S_0 D_0^T \quad [4.15]$$

The main innovation of LSA is the possibility of considering only the k greatest singular values in the matrix S_0 and reducing the other values to zero. The value of k is a pre-defined parameter: in general, values between 100 and 200 are used. The matrix of X is then approached by $X'=TSD^T$, where T is the matrix $t \times k$ with orthonormal columns, S is a positive diagonal matrix defined as $K \times K$ and D is the matrix $d \times K$ with orthonormal columns. Figure 4.33 presents the SVD of X, where the parts considered to obtain X' are shaded. To calculate the SVD of a matrix, most high-level programming languages have matrix calculation libraries that contain functions specifically dedicated to this purpose. The interest of the SVD here is not to reconstruct the original matrix that we already have, but rather to find an approximation of rank k in order to improve the search efficiency.

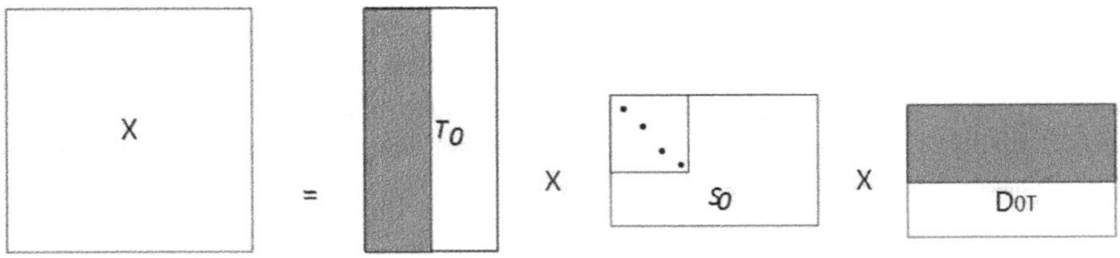

Figure 4.33. *Application of the SVD to a matrix X*

The last step consists of calculating the distance between the vector of the query (q) and the matrix of every document (d), the most common of which is the cosine distance that is calculated according to equation [4.16]. The cosine distance between two vectors is found by dividing the dot product by the multiplication of the absolute values of the two vectors:

$$\frac{q \cdot d}{|q||d|} \qquad [4.16]$$

To clarify the different steps of LSA processing, consider this example. Let's start by supposing that there is a collection of documents typed by a user (with a few errors) and the queries presented in Figure 4.34. For other detailed examples, see [LAN 98, GRO 04].

Doc$_{01}$ The transportation of vegetables is the most important.

Doc$_{02}$ Some fruit arrives soon in the big truck of fruit.

Doc$_{03}$ In one truck of vegetables.

Q$_{01}$ fruit truck

Q$_{02}$ vegetable transportation

Figure 4.34. *Collection of documents*

In the three micro-documents in our collection, the words do not have the same distinctive role. For example, the word *the* figures in two documents out of three and consequently only has a limited distinctive interest. On the other hand, the word *fruit* is only employed in the second document, and it is even repeated twice, which makes it very characteristic of this document.

Thus, from our collection of documents, we obtain the term–document matrix in Table 4.7. It should be noted that, given the extremely reduced size of our collection of texts and given the pedagogical objective, we have decided not to carry out lemmatization or the removal of empty words and we have used the number of occurrences of forms instead of, for example, the specificity values of terms like tf-idf.

Term	**Doc_{01}**	**Doc_{02}**	**Doc_{03}**	**Q_{01}**	**Q_{02}**
The	2	1	0	0	0
Transportation	1	0	0	0	1
Of	1	1	1	0	0
Vegetables	1	0	1	0	1
Is	1	0	0	0	0
Most	1	0	0	0	0
Important	1	0	0	0	0
Some	0	1	0	0	0
Fruit	0	2	0	1	0
Arrives	0	1	0	0	0
Soon	0	1	0	0	0
In	0	1	1	0	0
Truck	0	1	1	1	0
Big	0	1	0	0	0
One	0	0	1	0	0

Table 4.7. *Term–document matrix for our collection and the two queries*

The next step consists of calculating the SVD of the term–document matrix to find an approximation of it according to equation [4.15]. With an approximation factor of two (where k=2) for our example, we will only consider the first two columns of the matrix. Thus, we obtain the matrices in Figure 4.35 whose multiplication is approximately equal to the original matrix X2 (equation [4.15]).

$$T_2 = \begin{bmatrix} -0.47528302 & 0.38126428 \\ -0.13957648 & 0.29276679 \\ -0.41904627 & 0.07889814 \\ -0.2229162 & 0.28316743 \\ -0.13957648 & 0.29276679 \\ -0.13957648 & 0.29276679 \\ -0.13957648 & 0.29276679 \\ -0.19613007 & -0.20426929 \\ -0.39226014 & -0.40853857 \\ -0.19613007 & -0.20426929 \\ -0.19613007 & -0.20426929 \\ -0.27946979 & -0.21386864 \\ -0.27946979 & -0.21386864 \\ -0.19613007 & -0.20426929 \\ -0.08333972 & -0.00959935 \end{bmatrix} \quad S_2 = \begin{bmatrix} 3.92552161 & 0.00 \\ 0.00 & 2.80022548 \end{bmatrix}$$

$$D_2^T = \begin{bmatrix} -0.54791048 & -0.76991282 & -0.32715188 \\ 0.81981301 & -0.57200006 & -0.02688036 \end{bmatrix}$$

Figure 4.35. *Approximation factor of 2 of the SVD of X*

To integrate the queries into the process, we consider that queries are also documents as can be observed in Figure 4.33. From equation [4.15], we can infer equations [4.17] to calculate the vector of a document D and a query Q:

$$d = D^T T_2 S^{-1}$$

$$r = r^T T_2 S^{-1} \qquad [4.17]$$

To calculate the vector of Q01, which corresponds to the query (fruit truck) still with k = 2, in reduced space, we proceed as in Figure 4.36. By carrying out a similar processing for all of the documents, we obtain their vectors in reduced space. The vectors of the other documents are also given in Figure 4.36.

These vectors then serve to calculate the distances between the query and each of the documents according to equation [4.16]. It should be noted that these are directed distances (their value can be negative) and therefore d(a,b)=−d(b,a)), contrary to the general notion of distance, which can only be positive and for which d(a,b)=d(b,a).

$$Q01 = [0\ 0\ 0\ 0\ 0\ 0\ 0\ 0\ 1\ 0\ 0\ 0\ 1\ 0\ 0] \begin{bmatrix} -0.47528302 & 0.38126428 \\ -0.13957648 & 0.29276679 \\ -0.41904627 & 0.07889814 \\ -0.2229162 & 0.28316743 \\ -0.13957648 & 0.29276679 \\ -0.13957648 & 0.29276679 \\ -0.13957648 & 0.29276679 \\ -0.19613007 & -0.20426929 \\ -0.39226014 & -0.40853857 \\ -0.19613007 & -0.20426929 \\ -0.19613007 & -0.20426929 \\ -0.27946979 & -0.21386864 \\ -0.27946979 & -0.21386864 \\ -0.19613007 & -0.20426929 \\ -0.08333972 & -0.00959935 \end{bmatrix} \begin{bmatrix} \frac{1}{3.92552161} & 0.00 \\ 0.00 & \frac{1}{2.80022548} \end{bmatrix}$$

$$R_{01} = [-0.17457428 \quad -0.21922826]$$
$$R_{02} = [-0.09420735 \quad 0.20285925]$$
$$doc_{01} = [-0.5589752 \quad 0.80859255]$$
$$doc_{02} = [-0.78546074 \quad -0.56417132]$$
$$doc_{03} = [-0.33375851 \quad -0.02651246]$$

Figure 4.36. *Vector calculation of the queries and the documents in this collection*

Calculating the other distances in the same way and repeating the same process for the query R_{02} (vegetable transportation) gives the results in Table 4.8.

Document/Query	Q_{01}	Q_{02}
Doc_{01}	−0.28925511	0.98556838
Doc_{02}	0.96230936	−0.18701158
Doc_{03}	0.68292419	0.34805227

Table 4.8. *The distances between the queries and the documents in this collection*

Presented in a bidimensional space, this gives the vectors in Figure 4.37.

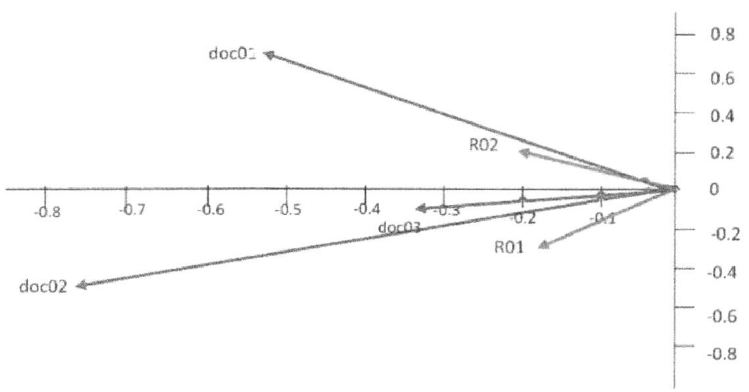

Figure 4.37. *The vectors corresponding to the texts in reduced space (k = 2). For a color version of this figure, see www.iste.co.uk/kurdi/language2.zip*

The order of the distances between the first query and the three documents is doc02, doc03, doc01. This order can be explained by the presence of the words *fruit* and *truck* in doc02 and the absence of these two words in doc01. Similarly, the order of similarity of the three documents in relation to the second query is doc01, doc03, doc02. This can be explained by the fact that the words *vegetable* and *transportation* occur in doc01 while only the word vegetable is used in doc03. Doc02, which does not contain any words in the query, is in the last place. One of the advantages of this method is that it is robust in the face of the variabilities that can exist in texts (typographic errors) and does not depend on a specific vocabulary.

Finally, it should be noted that, given the importance of information retrieval in today's world, many techniques from AI have been applied to this problem with varying degrees of success, including the popular Support Vector Machine (SVM) method [COR 95], neural networks and genetic algorithms [HSI 95].

4.4. Big Data (BD) and information extraction

With the democratization of the Internet, there was an explosion in the size of digital data. Gradually, there was a move from mostly formatted data (search results, information sites) toward more personal and sometimes intimate data. This large mass of data, impossible to exploit manually, required the emergence of a new discipline called *data science* that includes another even younger discipline called *Big Data* [RUS 14b, MAN 14, HUR 13].

4.4.1. *Structured, semi-structured and unstructured data*

BD applications are distinguished by the use of a variety of information resources that are available in various formats and types. Classically, computer systems use structured data. These data follow a strict framework typically defined in the form of a database table. These tables have relations between them that are defined through connections that are called foreign keys. The advantage of this framework is that there are theoretical foundations that make it possible to reduce data redundancy as much as possible through a standardization process as well as the existence of platforms, called Database Management Systems (DBMS), which guarantee rapid access to information even in cases of large databases. To manipulate this data, SQL (Structured Query Language) and its procedural derivatives such as TransactSQL and PL/SQL play an important role.

Semi-structured data have less rigid structures. They are notably represented in the form of XML documents. As shown in section 1.2.2.2, with the DTD, it is possible to define a general structure that can be instantiated in several possible forms.

Unstructured data are organized without constraints on the content (see section 4.3.1). Among others, it can consist of scientific, journalistic, personal or literary texts. Regarding formats, the data also vary considerably. Another important challenge worth noting regarding the data is that it consists of data distributed on several platforms on a network that can also be localized in distant geographical areas. This disposition, which is commonly called the cloud, has opened the door to new approaches.

4.4.2. Architectures of BD processing systems

BD are commonly considered to be an evolution of Business Intelligence. According to this approach, a company's data can all be gathered into a single space that is called a data warehouse where they are processed offline unlike databases where the data is modified and accessed online. Data warehouses are typically constructed according to the Extract Load Transform (ELT) approach (see Figure 4.38).

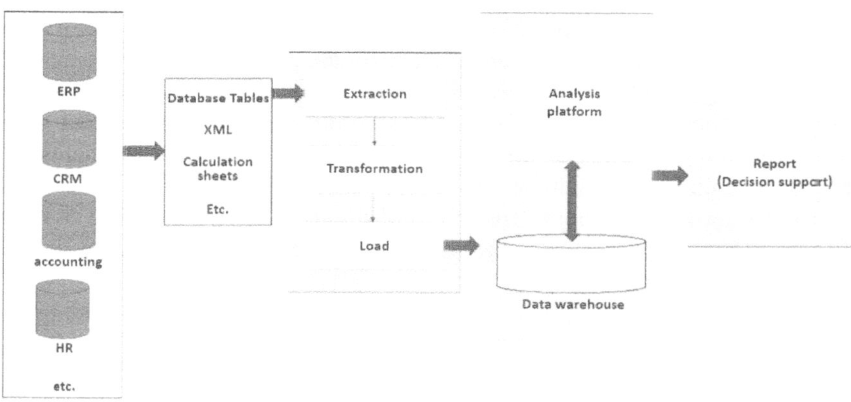

Figure 4.38. *ELT approach for data warehouses. For a color version of this figure, see www.iste.co.uk/kurdi/language2.zip*

The extraction, the most difficult phase, consists of collecting data from several systems. These data are typically relational databases with sometimes semi-structured data. The transformation itself applies a set of rules that aim to modify the data in order to guarantee its coherence in the warehouse. Every database designer can adopt a unique convention to represent information. For example, some use the letters *M* and *F* to indicate the genders masculine and feminine, respectively, while others prefer the digits 1 and 2 for the same task. Other differences that require a harmonization are the adoption of different units (weight, sizes, distances, etc.). The last phase consists of introducing the data in the data warehouse according to processes that vary considerably from one data warehouse to another. It is important to note that, as these three steps are purely technical and as they do not involve real scientific challenges, they are easily mastered and implemented in many companies.

Once the data is stored in a warehouse, *data mining* makes it possible to recognize patterns such as the products that customers buy together or in relation to a socio-economic profile in order to predict future behaviors. Classic algorithms like decision trees, neural networks and Bayesian networks are used to carry out this task (see [WIT 05, BER 04] for more details).

Because Big Data processing is a domain that is still in development, there is not yet one architecture that can be called a standard. Figure 4.39 presents a possible architecture for which there are many alternatives in the literature. The central element of this architecture is the parallel file system Hadoop, an essential support for parallel processing (implemented in clusters, groups of computers working in parallel on shared disk spaces), which has become essential to talk about BD. Initially constructed by Doug Cutting of Yahoo!, Hadoop is currently an open-source product distributed by the Apache foundation, notably known by its HTTP server. Two main components are at the core of Hadoop: the distributed file system and the MapReduce engine. Written in Java, Hadoop's distributed file system is a system designed to process distributed files in a portable way. This file system uses a TCP/IP communication layer and Remote Procedure Call (RPC) clients. The MapReduce engine is a parallel and distributed implementation of the data-processing algorithm, making it possible to distribute one calculation over several machines (the map phase) and aggregate the different results into one coherent whole (the reduce phase).

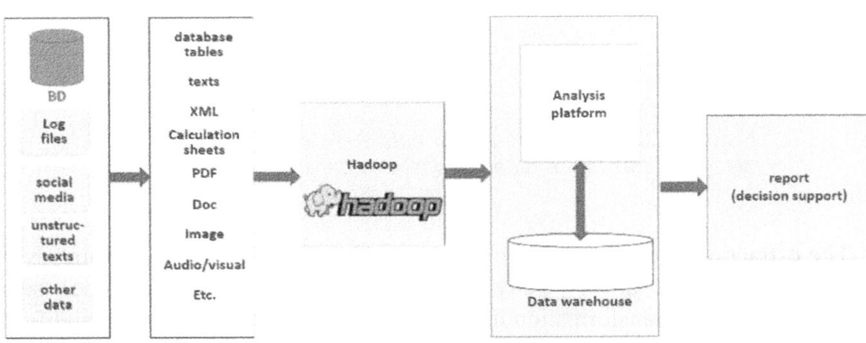

Figure 4.39. *BD processing architecture. For a color version of this figure, see www.iste.co.uk/kurdi/language2.zip*

The data are then analyzed statistically with pattern-matching and data-mining algorithms. Of course, given that a large quantity is available in text form, NLP is a fundamental part of the analysis process, especially to extract information.

4.4.3. *Role of NLP in BD processing*

Combining robustness and depth in the processing was the objective of many works in the NLP domain at the start of the 2000s, notably in the domains of human–machine dialogue and speech translation. The problem is that this occurred to the detriment of the independence of the applicative domain as these applications required recourse to deeper knowledge about the domain concerned. With the advent of the challenge of BD, gradually the priority given to the robustness and independence in the domain became essential again to the detriment of the depth of analysis. To a certain extent, BD allowed for the redeployment of existing tools and approaches but in a different way through new architectures. The new philosophy was to make the most of existing tools, techniques and models. The new approach to NLP could be seen as a rising exploration of the applicative possibilities of NLP: starting from the simplest applications and attempting to go as far as possible toward the ultimate objective, which is the construction of applications that satisfy the demands of the market.

However, as Big Data is a new form of input, it entailed changes to data-processing strategies. The abundance of data facilitated the use of statistical or, more generally, machine learning algorithms. In addition, this same abundance required the use of unsupervised or lightly supervised learning algorithms in order to reduce efforts related to annotating the data that are necessary for supervised learning approaches. This involves adapting these algorithms, which were initially designed to process thousands of cases, to process millions of cases. The parallelization required by environments constructed on Hadoop, even for high-level environments like the ESB (Enterprise Service Bus) TALEND[18] that offer an integrated graphic programming interface from the base code (java, Spark) to the project management, requires revising the design of NLP algorithms to adapt them to the MapReduce model.

Moreover, because the data is distributed in clouds, this requires adopting new strategies to process them. The distribution of data on different servers can also be accompanied by a topical distribution. For example, the data in one particular server can pertain only or mainly to the medical domain, while the data in other servers may concern biology or the environment. This requires algorithms that are capable of extracting general principles independent of the task while being able to merge the parameters related to the applicative domains distributed over different servers.

18 https://fr.talend.com

4.4.4. *Information extraction*

Information extraction is the extraction of structured information from unstructured information. It is a superficial approach that consists of finding partial information that is determined to be useful in a text or a data collection. The information extracted can concern security, transportation, urban planning, scientometrics, etc. Often, the outcome of the information extraction modules are analyzed statistically in order to recognize trends. A common example is the calculation of positive and negative opinions about a product, a politician, an institution, etc. [GRI 10]. Compared to information retrieval, information extraction involves extracting structured data from a structured query while the queries in an information retrieval system are more open. The common point remains that the search space is a collection of documents (see Figure 4.40).

Question answering systems can be seen as a variant of information extraction systems. The main difference is that these systems have an interface capable of analyzing questions input and a module capable of transforming the answer into an utterance. For more details about question answering systems, see [GUP 12, WEB 10] (see Figure 4.40).

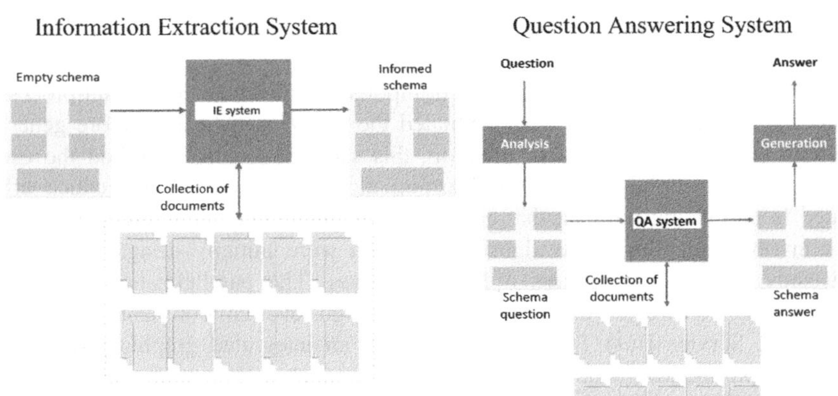

Figure 4.40. *DFD of an information extraction system and a question answering system*

As shown in Figure 4.40, the main difference between an information extraction system and a question answering system is the existence of two analysis and generation modules that are capable of processing open questions produced by users and generating responses in natural language, respectively. It should be noted that, in order to facilitate the comparison, in the case shown, the analysis and generation modules use schema-based semantic representation; however, in reality, other forms of semantic representations are also possible, such as representations based on different types of logic (modal logic[19], fuzzy logic[20], etc.) (see section 2.2.2) or even word embeddings on a large scale (see [MIK 13]).

4.4.4.1. *Extraction of named entities (NEs)*

The extraction of NEs consists of identifying and classifying proper nouns in pre-defined categories based on the needs of a given application. Typically, three basic categories are used: person, place and organization. In addition to these three large categories, sometimes emails, temporal expressions, measurement expressions like monetary amounts, distances or weight, bibliographic references, etc. are added. Without needing to create connections between the identified entities, entity extraction can be used as a module in a larger information extraction system. As demonstrated with the Prolex proper noun database presentation, proper nouns are an important part of a text, especially in journalistic and scientific domains, where they account for about 10% of words according to some studies. For an overview of works related to this question, see [NAD 07].

The extraction of NEs seems to be a simple task at first glance, but in practice there are many rhetorical phenomena, including notably metonymy, that make identifying them problematic. As an example, consider the set of sentences [4.18] (for more details, see the section on metonymy in Chapter 2).

Team vs. country	France won the World Cup. The World Cup was organized in France.	
Location vs. authority	The President of France officially resides at the Elysée. The Elysée declared that elections would not take place next month.	[4.18]
Company vs. product	He drinks Coca Cola every morning, he's crazy! Coca Cola just launched a new advertising campaign.	

19 https://mally.stanford.edu/notes.pdf
20 www.francky.me/doc/course/fuzzy_logic.pdf

According to the approach of Message Understanding Conference (MUC) evaluation campaigns, it could be considered that *France* is always a place but this does not seem to be an ideal solution. At the same time, it is not always possible to count on general ontological knowledge in order to distinguish the different uses of a term. The simplest method of processing NEs consists of using an appropriate dictionary like Prolex. Despite its speed and simplicity, this approach suffers from a major inconvenience, which is the updating required for the dictionary to reflect the current changes that, naturally, have an impact on the proper noun dictionaries in a much more substantial way than on ordinary dictionaries. Moreover, such a dictionary is not sufficient to resolve ambiguity, because its resolution requires using contextual information. Another approach consists of using regular expressions to consider internal and external indicators. Internal indicators are morphemes, which are part of the proper noun, but through their semantic nature, they indicate a location. Common examples of internal indicators are lake, river, airport, city, street and boulevard. This makes it possible to predict the role of the rest of the sequence like in the segments in italics in sentences [4.19]:

a) John lives on *Lexington Avenue*.

b) The *Saint Louis Lambert Airport* is one of the prettiest airports in the US. [4.19]

c) The *Riverside Park* is located along *Hudson river*.

d) The *city of Knoxville* is located in the *south of the US*.

External indicators are not part of the noun properly speaking but rather part of the cotext. They indicate that an NE will follow. As an example, consider the following indicators: in the south of, in the north of, beside and very close to.

Pattern-based methods, despite their simplicity, are very limited and do not make it possible to resolve many cases of ambiguity (e.g. John F. Kennedy, the airport or the person). This pushed researchers to adopt learning-based approaches like HMMs [BIK 97, BOU 02], SVM [TJO 03], naive, semi-supervised [PAS 08] or unsupervised [NAD 06] Bayesian networks.

As an example, consider the KnowItAll system developed by Oren Etzioni and his collaborators at the University of Washington [ETZ 05]. This system uses an unsupervised approach that is based on a three-step processing (see Figure 4.41).

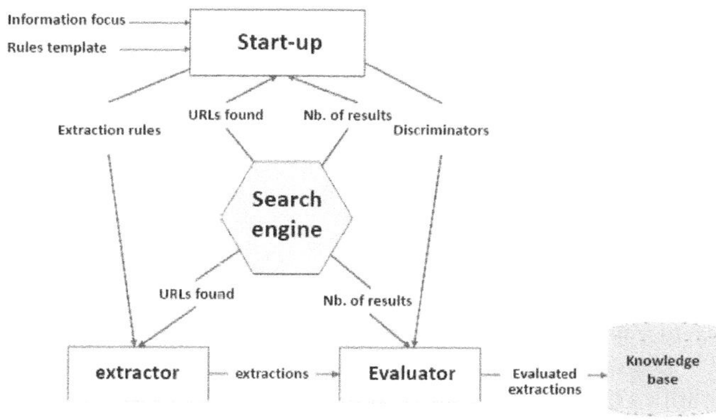

Figure 4.41. *DFD of KnowItAll*

First, the system proceeds to the start-up phase that consists of providing a set of predicates that represent classes of entities like movieActor, city and brand of car. Each of these predicates is associated with a set of labels that are the surface forms that can take an entity in an utterance. For example, consider the predicates for the domains of geography and films given in Figure 4.42.

Predicate: City Labels: city, metropolis, city-state Predicate: country Labels: country, nation Predicate: capitalOf (city, country) Relation label: capitalOf Class 1 label: city, metropolis, city-state Class 2 label: country, nation	Predicate: Film Labels: movie, feature film, medium-length film, short film Predicate movie actor: movieActor Labels: movie actor, movie star Predicate: starsIn(movieActor, Film) Relation label: starsIn, starIn Class 1 label: movie actor, movie star Class 2 label: movie, full-length film, medium-length film, short film

Figure 4.42. *Examples of predicates in the domains of geography and films*

Inspired by a Hearst pattern-based approach, the second step consists of labeling using eight generic patterns to produce candidates. The labels identified in the start-up step can serve to instantiate patterns with the appropriate classes. As an example, consider sentence [4.20]. In this sentence, the word city serves to indicate the class of entities that follow. Using the pattern "NP such as ListNP", we can analyze sentence [4.20] and propose four candidates for the class of city: Richmond, Charlottesville, Lynchburg and Roanoke.

> Our buses stop at cities like Richmond, Charlottesville, Lynchburg and Roanoke. [4.20]

Based on the Pointwise Mutual Information (PMI) algorithm developed by David Turney [TUR 01], the system then tests the plausibility of each hypothesis using the Web as a corpus. Thus, the system uses a search engine to bring up pages that contain entities that it intends to verify. For example, to verify that Lynchburg is the name of a city, the system seeks to verify that there is a mutual information (PMI) between Lynchburg and discriminating sequences (or discriminators) that indicate that it really is a city, such as: -is a city, the city of-, etc. Formally, the PMI score is calculated using equation [4.21]:

$$\mathrm{PMI}(I, D) = \frac{|\mathrm{Results}(D+I)|}{|\mathrm{Results}(I)|} \quad [4.21]$$

The PMI score is the number of results for a query that combines the discriminator D and the instance I divided by the number of results for the instance itself. To transform PMI scores into a co-occurrence probability, the system takes a set of *k* positive cases and *k* negative cases for each class and defines the threshold that serves to divide all of the cases. Then, another set of *k* positive cases and *k* negative cases is used to estimate the following probabilities:

$P(\mathrm{PMI} > \mathrm{threshold} \mid \mathrm{class})$, $P(\mathrm{PMI} \leq \mathrm{threshold} \mid \mathrm{class})$,

$P(\mathrm{PMI} > \mathrm{threshold} \mid \neg\mathrm{class})$, $P(\mathrm{PMI} \leq \mathrm{threshold} \mid \neg\mathrm{class})$,

It should be noted that in the experiments described, the value of *k* is equal to 10. The probabilities obtained are then used as input features for a naive Bayesian classifier according to equation [4.22], which shows the probability of a fact given the features $t_1, t_2, \ldots\ldots t_n$:

$$P(\phi|t_1, t_2, \ldots t_n) = \frac{P(\phi)\prod_i P(t_i|\phi)}{P(\phi)\prod_i P(t_i|\phi) + P(\neg\phi)\prod_i P(t_i|\neg\phi)} \quad [4.22]$$

As emphasized in [ETZ 05], the use of PMIs presents problems that merit further reflection, like rare data for some cases, even with the Web as a corpus, and the case of homonyms.

4.4.4.2. *Extraction of relations*

The relations between the entities in a text can be useful for many applications. For example, in a bibliographic information extraction application, it is essential to associate the articles with their authors and institutions. In medical applications, it is possible to extract relations between the genes that cause certain illnesses [CHU 06]. The extraction of relations can have the goal of the automatic construction of ontologies [WON 09].

The task of extracting relations can be described as the construction, from raw texts, of predicates, whose diagram is: relation (e_1, e_2, …e_n), where *e* represents an entity cited in the text. Here are a few examples:

Emmanuel Macron is the president of France.	president (Emmanuel Macron, France)
Ouagadougou is the capital of the African Republic of Burkina Faso.	capital (Ouagadougou, Burkina)
The discus is a very pretty fish that lives in the Amazon.	lives (discus, Amazon)
Nadine is the mother of Celine and Amandine.	mother (Nadine, Celine, Amandine)

The simplest approaches are based on pattern recognition. [HEA 92] proposed five patterns to extract hyponyms (see Table 4.9).

Patterns	**Examples**
Y such as X_1, X_2, ..., X_n	A sports car like a Ferrari, Bugatti or Porsche.
X_1, X_2, ..., X_n and/or/among others Y	Car, train, plane among other modes of transportation
Y, including X_1, X_2, …, X_n	Combat sports, including karate, judo and boxing
Y, in particular X_1, X_2, …, X_n	Planes, in particular the F22 and F35.
Some Y such as X_1, X_2, …, X_n	Programmers like John, Patrick and Will

Table 4.9. *Hearst's extraction patterns for hyponyms*

The idea behind supervised approaches is to manually annotate a corpus and automatically extract patterns and rules. The general diagram of this process is shown in Figure 4.43.

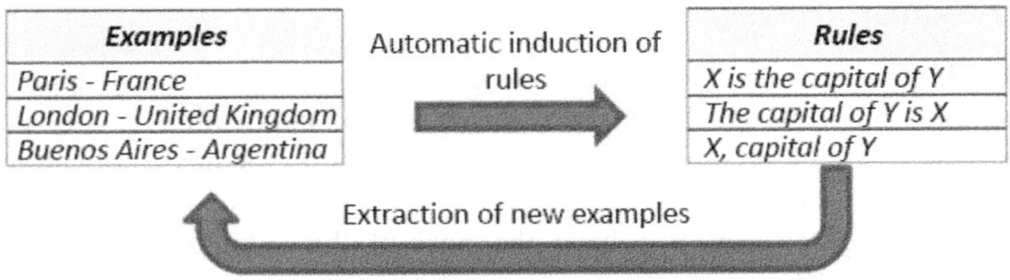

Figure 4.43. *Induction of rules from examples*

To make the process of writing rules less demanding, it is possible to start by annotating a certain number of cases manually in order to start the system of automatic rule induction. The system obtained can then be used to annotate new cases. The idea is to define a confidence threshold in order to select only cases annotated by the system whose confidence score is greater than the threshold. Over the course of the system's development, its performance improves and the percentage of cases annotated with an acceptable score increases until a maximum is reached (a plateau on the performance curve), where the system's performance stagnates even after the addition of new learning cases.

In the literature, two learning-based approaches are opposed (see [BAC 07] for an overview): feature methods and kernel methods [LOD 02]. In feature-based approaches, an input sequence analysis is used to attempt to identify the disposition patterns of the entities that are found in this sequence (see, for example, [KAM 04]), for example, the number of words that separate two entities and the types of entities (person, place or institution). Some approaches go further by using dependency trees for a sentence. For example, to extract relations from the sentence *John invented a new steam engine in his garage*, we can use a decision tree like the one in Figure 4.44.

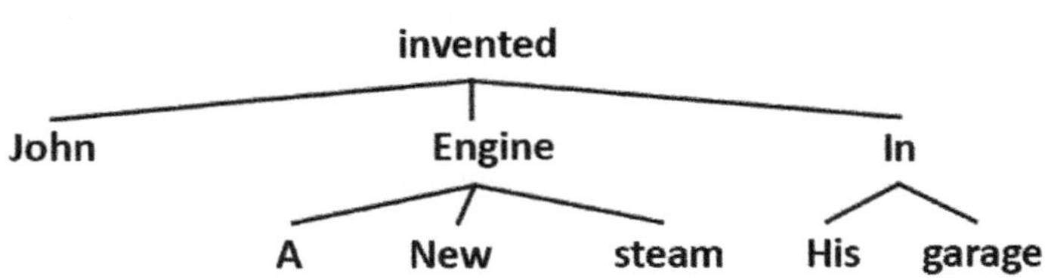

Figure 4.44. *Dependency tree to extract relations*

In general, a set of features is the result of a process of trial and error. More formally, a sentence S with n entities and k words is presented in this way: $F = w_1, w_2, e_1, w_i, e_2, \ldots, w_l, \ldots, e_n, \ldots, w_{k-1}, w_k$, where $k > n \geq i$. The sentence thus represented is analyzed by a classification function f_R that provides two defined outputs according to these criteria:

$$f_R(P) = \begin{cases} +1 \; \textit{if the entities } e_1, e_2, \ldots e_n \textit{ are related by the relation } R. \\ -1 \; \textit{if the entities } e_1, e_2, \ldots e_n \textit{ are not related by the relation } R. \end{cases}$$

The classification function can be realized by any classification algorithm, for example, a variation of an entropy-based statistical algorithm.

Kernel-based approaches use structural representations with similarity scores to guide the classifier and thereby considerably reduce problems related to the choice of similarity features. The principle of these approaches is to use a function K(X, Y) that defines the similarity relation between the objects X and Y. For example, when considering the level of character sequences, the similarity of two sequences increases with the number of sub-sequences shared between these two sequences. Thus, the similarity between the sequences is sim(livre, livrent) > sim(livre, livarot)[21].

The integration of supplementary knowledge resources such as parts of speech and syntactic analysis in segments or dependency trees has also been used to complete this task [ZEL 03, CUL 04, ZHA 05]. For example, to calculate the similarity between two analysis trees, a relatively simple algorithm can be used (see Figure 4.45).

```
Compare the root node features,
        If the two roots are identical, then add 1 to the identity score.
        If not, zero
Repeat the same process recursively to the child nodes
```

Figure 4.45. *Algorithm to calculate tree similarity*

Given the cost of preparing data in supervised learning-based approaches, researchers gradually began to explore semi-supervised and unsupervised learning approaches [BUN 07, BOL 11]. Semi-supervised approaches are a compromise between the minimal effort of unsupervised approaches and the performance of supervised approaches. The principle of semi-supervised learning consists of annotating a limited number of cases to avoid the effort required to annotate an entire corpus and using them to construct groups of cases that are called *sentence*

21 In French, *livre* is book, *livrent* is deliver and *livarot* is a type of cheese.

bags. A sentence bag is constructed like this: first, some entities that are related or not by a known semantic relation are identified. For example, [BUN 07] considers the case of companies acquired by multi-nationals (see examples in Figure 4.46).

CorpAcquired(Google, YouTube)
CorpAcquired(Adobe Systems, Macromedia)
CorpAcquired(Viacom, DreamWorks)
CorpAcquired(Novartis, Eon Labs)
not(CorpAcquired(Yahoo, Microsoft))
not(CorpAcquired(Pfizer, Teva))

Figure 4.46. *Pairs of acquisition by multi-nationals*

Then, a search is done on the Web using the two entities in each predicate as keywords. The results that the search engine provides are verified in a large number of pages but not necessarily all pages. Moreover, some pages may discuss the acquisition of YouTube by Google more or less explicitly while others may provide information related to these two entities without necessarily evoking their mutual relationships. It can also be assumed that the relation between these two entities will not be verified in practically all of the results where it is not a proven fact. More concretely, it can reasonably be assumed that the number of pages that a search engine will present that discuss the acquisition of Yahoo by Google is zero. Therefore, the number of pages serves as a factuality indicator for the supposed relation between the entities. The set of results for a pair of entities is called a *case bag*. Some examples of search results with the acquisition pairs are given in Figure 4.47.

+/S1: Search engine giant Google has bought video sharing website YouTube in a controversial $1.6 billion deal.

–/S2: The companies will merge Google's search expertise with YouTube's video expertise, pushing what executives believe is a hot emerging market of video offered over the Internet.

+/S3: Google has acquired social media company YouTube for $1.65 billion in a stock-for-stock transaction as announced by Google Inc. on October 9, 2006.

+/S4: Drug giant Pfizer Inc. has reached an agreement to buy the private biotechnology firm Rinat Neuroscience Corp., the companies announced Thursday.

–/S5: He has also received consulting fees from Alpharma, Eli Lilly and Company, Pfizer, Wyeth Pharmaceuticals, Rinat Neuroscience, Elan Pharmaceuticals and Forest Laboratories.

Figure 4.47. *Examples of search results*

Finally, it is important to note that the way to proceed corresponds to the Multiple Instance Learning (MIL) paradigm that was initially proposed by [DIE 97] (see [AMO 13] for a general presentation of this paradigm).

4.4.4.3. *Event extraction*

Event extraction was one of the first applications in the domain of information extraction. The events concerned were as varied as terrorist attempts, road accidents and announcements. The events could be extracted from written texts or telephone conversations like in [BOU 02]. According to the terminology of the Message Understanding Conference (MUC), each event is characterized by a scenario template that contains a set of slots that must be instantiated.

Just like the extraction of named entities and relations, supervised [SOD 99, YAN 00], semi-supervised [SEK 07] and unsupervised [SHI 06] learning approaches were used for the extraction. Several rule-based and pattern-based systems were created like the Fastus system by [APP 93] and the REES system by [AON 00]. For example, to extract events, [AON 00] proposed an architecture that conducts a processing in three main steps: a tagging module, a coreference resolution module and a pattern generation module (see Figure 4.48).

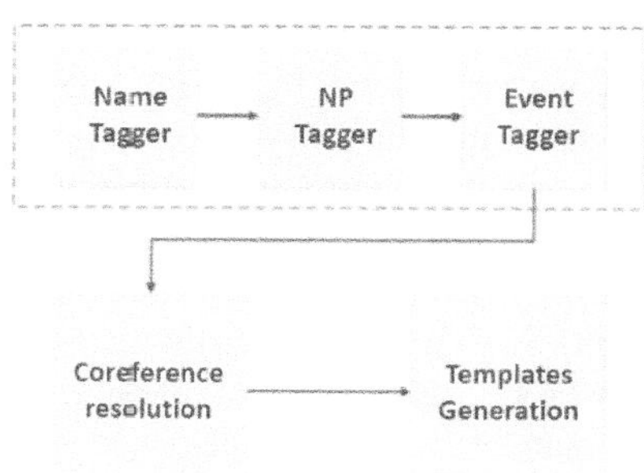

Figure 4.48. *DFD of the REES system*

The tagging module itself conducts a three-step processing: tagging of names, Noun Phrases (NP) and events. These three components use the same processing engine but different templates for each component. The coreference module then proceeds with labeling the nominal groups that correspond to organizations, people, places, etc. The template generation module is based on declarative rules. It results in a template whose format is compatible with the MUC recommendations, as shown in Figure 4.49.

<ATTACK TARGET-AP8804160078-12>: = i
TYPE: CONFLICT
SUBTYPE: ATTACK TARGET
ATTACKER: [TE for *"an Iraqi warplane"]*
TARGET: [TE for *"the frigate Stark"]*
WEAPON: [TE for *"missiles"]*
TIME: "May *17, 1987"*
PLACE: [TE for *"the gulf"]*
COMMENT: *"attacked"*

Figure 4.49. *Example of an event template*

Then, REES concatenates the templates that are associated with the same event. This occurs using a set of declarative rules designed based on world knowledge. For example, as the birth and death of a person are unique events, there is a rule that makes it possible to concatenate templates relating to such events when they have the same subject.

4.4.4.4. *Sentiment analysis and opinion mining*

While working on interactive systems like virtual agents or robots, researchers realized that some affective information processing is inevitable because it represents a very important implicit and subjective part of all human communication. For example, a virtual guide in an airport would be too monotonic and therefore not very useful if it could not detect the emotions of its interlocutor and adapt its tone [SLO 81, SMI 10, OCH 12]. In terms of generating answers on the part of the agent, many processes also play a role in expressing a sentiment or an opinion in an appropriate manner: lexical choices, syntactic constructions, prosody, etc. These elements play an important role in controlling the interaction, for example, by making it possible to prolong the dialogue or by participating in managing the focus (see, for example, [JAN 16]).

Similar applications have also been created to extract opinions expressed in a text or a forum, to observe investment tendencies and to detect criminal or terrorist communications. This is particularly important to automatically survey opinions about a given commercial product or even a political question. The popularity of social media like Facebook and Twitter has made the use of such practices on texts produced by users of these media particularly attractive [RUS 14a].

It should be noted that, given the constraints on the form as well as the content of these media, processing emotions is not as easy. Processing messages on Twitter is easier, given the concise nature of these messages and their direct character [PAK 12]. Processing product reviews is also facilitated by the direct aspect of these reviews and by the relatively minimal quantity of irrelevant information. On the other hand, blogs and forums are more difficult because they can address a number of important questions, they can make implicit or explicit comparisons, and there are also many digressions and sarcastic comments.

Currently, there is an abundance of literature concerning these questions under various names. For example, this includes opinion mining, opinion extraction, emotion analysis or extraction, subjectivity analysis, affection analysis and emotion detection. As noted in [RIL 06], the domain of subjectivity is quite vast and includes allegations, opinions, desires, beliefs, suspicions and speculations. Among these definitions, the notion of the opinion seems particularly pertinent. An opinion is a sentiment, attitude, emotion, an evaluation of an entity or a topic, or an aspect of these [KIM 04, LIU 11]. An opinion can be positive, negative or neutral (see, for example, [4.23] for a positive opinion).

Paul thinks that “Le Monde” is the best newspaper that covers political life in France. [4.23]

This is called the semantic orientation of the opinion, or the polarity of the opinion. Formally, an opinion can be expressed by a quintuplet (e_j, a_{jk}, so_{ijkl}, s_i, t_l), where [LIU 11]:

– e_j is the entity intended by the opinion;

– a_{jk} is an aspect or a feature of the entity e_j;

– so_{ijkl} is the value of an opinion of a source s_i on a feature a_{jk} of the entity e_j at a moment t_l. This value can be positive, negative or neutral;

– s_i is the source of the opinion;

– t_l is the time of expression of this opinion.

It consists of a generic definition that can apply to the political, commercial, social, or intellectual domains, and so on. It should also be noted that the correlations between these five elements are rather important: ideally, all five elements should be there to maximize the utility of the analysis. An opinion without an intended entity or its features has only a little importance. If we analyze the opinions expressed in sentence [4.23], we find the following elements:

– the entity intended by the opinion (ej): the newspaper “Le Monde”;

– aspect (ajk): political;

– value (soijkl): best/positive;

– source (si): Paul;

– time (si): present.

Since processing sentiments is inherently connected to subjectivities, it is not always possible to decide which sentiment provokes a given sentence in an absolute way. Consider sentence [4.24]. The sentiment that provokes this sentence could be the fear of being contaminated for someone who lives in Africa or sadness for someone who lives outside of Africa. Moreover, an African whose family is vaccinated against cholera will have a different sentiment compared to someone whose family is not vaccinated. To solve the problem of relative stance, in the early work, researchers simply avoided considering this variety of angles about sentiments and adopted a single perspective that they judged to be standard or representative of the majority of people. However, recent works have begun to consider the relative stance of the author who expressed the opinion (see, for example, [RAJ 16, EBR 16 and JAN 16]).

Cholera is devastating Africa this year. [4.24]

Psychologists define emotions as complex states of sensations that give rise to psychic or physiological changes that affect thought and behavior. Emotions are associated with a fairly important level of phenomena like temperament, personality, humor and motivation.

Historically, the study of emotions started by at least the 19th Century, notably with the works of William James in the domain of psychology. Since then, several theoretical models have been proposed. Some models, known as dimensional models, consider that emotional states are connected by a set of dimensions shared between all emotions. They also include intensity parameters. From a cognitive perspective, these models postulate the existence of a neurophysiological system that is centrally responsible for all affective states. This contradicts the point of view supported by basic emotion theories, which consider that emotions emanate from different neurophysiological systems. Despite the objectives of cultural or physiological universalism, several psychologists and anthropologists find that it is not possible to define a universal framework to classify emotions. It is known that some cultures encourage the expression of emotions while others judge such expression to be unacceptable. The divergence between the starting postulations is reflected even more significantly in the models, a few of which will be presented here.

The semantic differential model proposed by Osgood focuses on scaling the connotations of objects, events and concepts [OSG 57]. These connotations are used to derive an attitude with regard to an event, an object or a concept. After having studied several semantic differential scales, Osgood and his colleagues succeeded in identifying three recurrent attitudes to evaluate words. These consist of evaluating with the pair of adjectives: good/bad, the power: strong/weak and the activity: active/passive. They also showed in a study of about 10 cultures that these three dimensions are universally present throughout the studied cultures [OSG 75].

In turn, the categorical model of Ekman starts from the perspective that there are six discrete categories of sentiments that are expressed by words: anger, disgust, fear, joy, sadness and surprise. As he considered that these sentiments are universally translated by distinct facial expressions, Ekman concluded that these sentiments have a biological origin. Although these categories have been adopted by many works in the domain of automatic processing of sentiments, some researchers have postulated that these categories are dependent on the domain of application [DME 07].

The circumplex model was initially developed by [RUS 80]. This model considers that emotions are distributed in a circular bidimensional space with arousal and valence values, where the arousal represents the vertical axis while the valence represents the horizontal axis (see Figure 4.50). The point of intersection of the two axes is the center of the virtual circle of emotions and represents a neutral state. Thus, the different emotions can have different values of emotions or affections whose value is zero.

Figure 4.50. *The representation of emotions according to Russell's circumplex model*

Finally, the cubic model developed by Lövheim is distinguished by its biological inspiration because it relates the rates of three neurotransmitters, which are dopamine, noradrenaline and serotonin, with the eight basic emotions identified by Silvan Tomkins [LÖV 12]. Each of these emotions is associated with these neurotransmitters in varying degrees, as shown in Table 4.10.

Base emotion	serotonin	dopamine	Noradrenaline
Shame/humiliation	Low	Low	Low
Distress/Anguish	Low	Low	High
Fear/Terror	Low	High	Low
Anger/Rage	Low	High	High
Contempt/disgust	High	Low	Low
Surprise	High	Low	High
Pleasure/Joy	High	High	Low
Interest/excitation	High	High	High

Table 4.10. *The basic emotions and corresponding neurotransmitter rates*

The direct connection between neurotransmitters and emotions makes it possible to give biological explanations for emotions. For example, depression can be seen as a situation that causes a decrease in serotonin, thus limiting the possible emotions to Shame/Humiliation, Distress/Anguish, Fear/Terror and Anger/Rage. An inverse vision of this state would be to consider that it is impossible for someone with depression to have the emotions of surprise, joy or interest, which require higher levels of serotonin.

Tridimensional, Lövheim's model places Tomkin's eight basic emotions in the corners of a cube and uses the following three axes: noradrenaline, serotonin and dopamine.

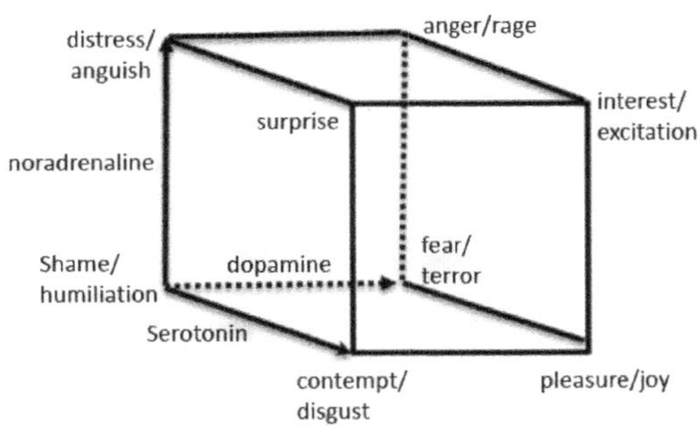

Figure 4.51. *Lövheim's cube of emotions*

The advantage of this classification is that it offers a way to account for differences between emotions that it is not possible to distinguish with bidimensional models.

Finally, the Hourglass model developed by Cambria is worth mentioning as well. It is a model based on the works of Plutshik on human emotions [PLU 01]. The fundamental idea of this model is that the mind is made of several independent sources and that emotional states result from the activation or deactivation of these resources [MIN 07]. Each of these resource combinations changes the brain's activities, thereby affecting our way of thinking. For example, the state of anger seems to select resources that allow for reacting quickly and with more force while suppressing other resources that cause a person to act cautiously. This idea is also confirmed by the results of research in the domain of cerebral imaging that show that a different pattern of cerebral activity accompanies each emotion (see [CAM 12, CAM 16] for more details about this model).

An emotion can be expressed using explicit linguistic means such as the use of verbs and expressions of opinions have extremely varied forms. For example, in English, this includes expressions like *I think, I believe, I consider, it seems to me, in my opinion* and *I am of the opinion*. Verbs that serve to express taste can also serve to express an opinion. In English, this includes following verbs like *cherish, adore, love, please, appreciate, detest, displease, prefer* and *be favorable*. However, the use of an explicit element of opinion expression is not a necessary condition for expressing an opinion. Opinions are very often expressed implicitly, and the use of a linguistic indicator only has the function of underscoring the subjective character of the opinion or, in certain cases, highlighting the person uttering it.

To express an opinion, three forms of expression can be distinguished: direct expression, indirect expression and comparative opinion.

A direct opinion consists of an explicit qualification of an entity, a topic or an action. This qualification can take the form of a qualificative adjective or epithet, or a lexical choice (see examples in Table 4.11).

Opinions	Type of qualification
This software is really efficient.	With an adjective
In this racer, it only takes an hour to get to Paris.	With a noun that has a positive connotation
Jules' company is just getting by after last year's losses.	With a verb that has a negative connotation

Table 4.11. *Examples of opinions with different types of qualifications*

Indirect opinions, in turn, are often formulated as the logical consequence of an action or a means (use of an entity), as in sentence [4.25]:

After using this detergent, Adeline's dishes shone like diamonds. [4.25]

Comparative opinions are used to compare entities, ideas or topics to express preferences. Linguistically, this occurs using comparatives and superlatives. Again, the comparison can be explicit or implicit as in [4.26a] and [4.26b], respectively:

a) Ferraris are less practical than Fords.

b) The left is the political current that defends workers' rights the most (implicitly more than the right and the center). [4.26]

To process the lexical level of opinions, an extension of the WordNet base was created at the IRST Institute in Italy to equip definitions of words in this base with an affective value. It is called WordNetAffect and it covers the entire lexicon of the English language [STR 04]. The words are classified according to the approach of [ORT 87] given in Table 4.12. The words are annotated with a polarity (positive or negative) as well as degrees to designate the force of an emotion. The 2004 version of WordNetAffect includes 1,314 synsets and 3,340 words. Several other projects intended to create lexical databases or dictionaries with emotions have been created. Each of these projects is distinguished by its method of annotating words (automatic, manual or semi-automatic), its size or taxonomy adopted (see, for example, [KAM 04b] and SentiWordNet[22] [ESU 06]).

Category	Examples of terms in English and French
Emotion	Anger – fear
Cognitive state	Doubt – dazed
Feature	Aggressiveness – competitive
Behavior	Cry – offense
Attitude	Skepticism – intolerance
Emotion	Pleasure – feel

Table 4.12. *Categories of terms in WordNetAffect [STR 04]*

22 http://sentiwordnet.isti.cnr.it

In terms of syntax, two syntactic forms directly affect the expression of sentiments and emotions: negation and dislocation. Negation makes it possible to inverse the polarity of an opinion or to restrict this opinion to a precise entity (see sentences [4.27]):

a) I do not like cream cheese. (Negative opinion)

b) She only likes white roses. (Restriction) [4.27]

Another syntactic construction that affects the expression of emotions is dislocation. It consists of displacing a phrase in order to highlight it (see Table 4.13). The focus attracts the attention of several opinion elements and therefore has various functions. Naturally, a good module of opinion extraction should consider these syntactic variations in order to be able to process the opinions correctly. Moreover, several works concern the role of the discursive structure in the expression of opinions, notably including that of [POL 04c], that studied the effect of discursive connectors on the polarity of an opinion. Consider sentences [4.28]:

Michael is good at networks. (Positive opinion)

Although Michael is good at networks, he is not very good at algorithms and programming. (Negative opinion) [4.28]

Sentences	Syntactic role of the element in focus	Opinion element in focus
It's <u>this song</u> that I like.	Agent	Intended entity
It is Jacqueline's <u>dress</u> that I adore.	Patient	Intended entity
It's <u>John</u> who likes red roses.	Agent	Source of the opinion
With much recognition and gratitude, he accepted your gift.	Prepositional group expressing the manner	Opinion

Table 4.13. *Relation between syntactic structures and the elements of an opinion*

The works of [VOL 07] highlighted the role of the summative evaluation of an entity that comes at the end to summarize the previous evaluations that concern several aspects of an entity or a topic. These researchers showed that attributing a more significant weight to these evaluations made it possible to improve the analysis results. Consider passage [4.29]. The evaluation given includes several elements that can be judged to be positive (new, specialized, effective service, rich menu), but the location of negative elements at the end gives them a conclusion value and attributes a globally negative aspect to the evaluation. Sometimes, explicit markers can be used to indicate the conclusion such as *in short*, *in sum* and *in summary*. Finally, it is also useful to mention that a large number of annotated linguistic resources are available for researchers (see [PAN 08] for a review of these resources):

> Located in a bright new building just steps from the Atlanta Marta train station, a restaurant specialized in Mexican cuisine recently opened its doors in our city. The service is very fast and the menu is fairly large. However, the lack of basic services like parking and its location close to the train station as well as music playing too loudly breaks the ambiance and brings to mind fast food restaurants. [4.29]

Subjectivity extraction is a very promising domain from a research point of view, especially from an industrial applications perspective. Making a detailed presentation of all of the applications is far beyond the scope of the present section, so this text will be limited to quickly presenting the vectorial approach (sometimes called categorical) that is one of the most commonly adopted approaches. Very similar to the approach of the same name in information retrieval, the analysis according to this approach occurs in four main steps (see Figure 4.52).

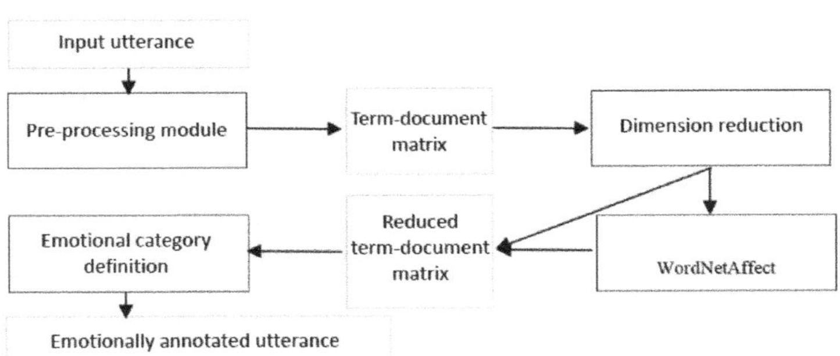

Figure 4.52. *DFD of a categorical system for extracting subjectivities*

First, the most characteristic elements in the document are identified. Many terms in a text are not useful for its categorization because they are shared by a large number of texts in diverse categories (see section 3.1.8). Another disadvantage is the computational cost related to processing all of the words in a text, whose numbers can be quite considerable. To overcome these two problems, emotion-processing systems, like information retrieval systems, proceed to a pre-processing phase whose role is to morphosyntactically annotate words, filter non-distinctive words and create a matrix of all significant words in a document (see section 4.3.2 for more details). The reduction in dimensions occurs through latent semantic indexing whose role is to break down the matrix obtained with the Singular Value Decomposition (SVD) process. The WordNetAffect module in turn enriches the simplified matrix with information related to emotions. Finally, the cosine distance is used to calculate the distance between the vector obtained and the emotional vectors.

Conclusion

This book provided a broad overview of NLP, which is a young discipline, but one that has reached a substantial applicative maturity, especially during this decade. The recent success of NLP, particularly in applicative domains, is due to both technological progress and the current historical circumstances.

With regard to technology, the Internet has become an integral part of society. On a daily basis, billions of people across the world use search engines available on the Web for personal and professional reasons. Opinion mining and emotion analysis have become essential for target marketing specialists as well as communication directors for political or artistic personalities. Similar tools have found a place within management software, for example, in the context of decision support systems. Such tools make it possible to survey the opinions of workers in a company about a given question by analyzing the content of electronic messages that circulate in public spaces on the Internet and social media. Some go so far as to analyze the morale of employees using similar tools, although this practice is not without ethical concerns and, as with any technology, it is not the tool itself that should be questioned, but the intention of the person wielding it.

With regard to the historical circumstances, in addition to the creation of multilingual political entities like the European Union, the world in general is increasingly connected, hence the need for multilingual tools that can respond to the growing demand for communication between people who do not share a common language. This makes NLP applications like machine translation and multilingual databases even more important for the future.

This book presented the concepts and works from the NLP domain in four branches: the lexicon and knowledge, semantics, discourse and applications. The first chapter, after a brief introduction to the basic concepts of lexical semantics, presented several data coding and exchange standards as well as the main dictionary writing systems. It also summarized several knowledge and ontology representation paradigms. The second chapter presented the concepts of formal and combinatory semantics. The third chapter presented discourse analysis from the perspective of linguistic models as well as segmentation and discourse interpretation applications. The fourth chapter presented several aspects related to NLP systems, such as their software architectures, in relation to different cognitive models, as well as the various approaches to evaluate them. The main NLP applications, notably including machine translation, information retrieval and information extraction from Big Data, were also presented.

NLP is an interdisciplinary domain *par excellence*. It draws its sources from a large number of related disciplines like computer science, AI, linguistics, cognitive psychology and neuroscience. Given this interdisciplinarity, the future of NLP will certainly be affected by developments in these connected domains.

Computer science is witnessing the emergence of new computational paradigms, such as quantum computing, whose realization makes it possible to extend the applicative field of NLP by allowing the large-scale deployment of applications known for their computational cost, like automatic speech translation. In addition, this makes it possible to explore new algorithmic approaches based on parallel processing, whose development has been hampered by the current computational cost.

Slow but consistent progress has been realized in modeling human cognition, notably due to new cerebral imaging technologies. Similarly, many fundamental works in the domain of cognitive engineering are striving to develop hybrid biological and electronic systems, combining natural and artificial neural networks or Brain Computer Interface BCI (see [CLE 16a, CLE 16b]). These works are paving the way for new approaches to NLP in terms of architectures and algorithms.

Finally, although progress on linguistic theories has been somewhat overshadowed by the mass of technological works and developments associating computing and linguistics (general and formal), it has now definitively branched out from the dominance of English. Researchers now have means that are far greater than those in the past to address languages that were previously neglected. Aside from the obvious practical advantage of this tendency for multilingual applications, I believe that, in time, this progress will encourage the emergence of new linguistic models that will offer a better consideration of the universal and the particular. The relation between general knowledge and linguistic knowledge is another domain that could see phenomenal progress. The resulting linguistic models will foster future generations of NLP applications, which will become more modular and consequently less dependent on any particular language.

Bibliography

[ABE 03] ABEILLÉ A., CLÉMENT L., TOUSSENEL F., "Building a treebank for French", in ABEILLÉ A. (ed.), *Treebanks*, Kluwer, Dordrecht, 2003.

[ACL 16] ACL, *Proceedings of the First Conference on Machine Translation (WMT),* Berlin, Germany, available at: http://www.statmt.org/wmt16/book.pdf, 7–12 August 2016.

[ADA 92] ADAM J.-M., *Les textes : types et prototypes*, Nathan, Paris, 1992.

[ADA 01] ADAM J.-M., "Types de textes ou genres de discours ? Comment classer les textes qui disent de et comment faire ?", *Langages*, vol. 35, no. 141. pp. 10–27, available at: http://www.persee.fr/web/revues/home/prescript/article/lgge_0458-726X_2001_num_35_141_872, 2001.

[ADD 98] ADDA G., LECOMTE J., MARIANI J. *et al.*, "The GRACE French Part-of-Speech tagging evaluation task", *The First International Conference on Language Resources and Evaluation (LREC)*, vol. 1, ELDA, Granada, pp. 433–441, May 1998.

[AGR 98] AGRAWAL R., GEHRKE J., GUNOPULOS D. *et al.*, "Automatic subspace clustering of high dimensional data for data mining applications", *SIGMOD '98 Proceedings of the 1998 ACM SIGMOD International Conference on Management of Data*, pp. 94–105, 1998.

[ALB 08] ALBACETE R.O., LÉON J., JORGE BOTANA G., "Using latent semantic analysis vs. human judgments assessing short summaries in expository texts", available at: http://www.researchgate.net/publication/267836513_Using_Latent_Semantic_Analysis_vs._Human_Judgements_assessing_short_summaries_in_expository_texts, 2008.

[ALL 95] ALLEN P.J., The train 93 dialogs, TRAINS Technical note94-2, The University of Rochester Computer Science Department, March 1995.

[ALL 01] ALLEN C., HAND M.L., *Logic Primer,* MIT Press, Cambridge, 2001.

[ALP 66] ALPAC, Languages and machines: computers in translation and linguistics, A report by the Automatic Language Processing Advisory Committee, Division of Behavioral Sciences, National Academy of Sciences, National Research Council, Washington, DC, National Academy of Sciences, National Research Council, Publication 1416, 1966.

[ALT 99] ALTMANN G.T.M., "Thematic role assignment in context", *Journal of Memory and Language*, vol. 41, pp. 124–145, 1999.

[AMI 13] AMINI M.-R., GAUSSIER E., *Recherche d'information : Applications, modèles et algorithmes, Applications, modèles et algorithmes – Fouille de données, décisionnel et big data*, Eyrolles, Paris, 2013.

[AMO 13] AMORES J., "Multiple instance classification: review, taxonomy and comparative study", *Artificial Intelligence*, vol. 201, pp. 81–105, 2013.

[AMS 06] AMSILI P., "Logique du premier ordre : une introduction aux linguists", available at: www.linguist.univ-paris-diderot.fr/~amsili/Ens06/mainBx.pdf, 2006.

[AND 77] ANDERSON J.M., "On case grammar", *Journal of Linguistics*, vol. 17. no. 2, pp. 374–378, 1977.

[AND 85] ANDREWS A., "Major functions of noun phrase", in SHOPEN T. (ed.), *Language Typology and Syntactic Description*, vol. 1, Cambridge University Press, Cambridge, 1985.

[AND 07] ANDREWS N.O., FOX E.A., Recent developments in document clustering, VirginiaTech Technical Report, available at: http://eprints.cs.vt.edu/archive/00001000/01/docclust.pdf, 2007.

[ANT 00] ANTOINE J.-Y., SIROUX J., CAELEN J. *et al.*, "Obtaining predictive results with an objective evaluation of spoken dialogue systems: experiments with the DCR assessment paradigm", *LREC'2000*, Athens, Greece, 2000.

[ANT 02a] ANTOINE J.-Y., LETELLIER-ZARSHENAS S., NICOLAS P. *et al.*, "Corpus OTG et ECOLE_MASSY : vers la constitution d'une collection de corpus francophones de dialogue oral diffusés librement", *Actes TALN'2002*, Nancy, France, pp. 319–324, June 2002.

[ANT 02b] ANTOINE J.-Y., BOUSQUET-VERNHETTES C., GOULIAN J. *et al.*, "Predictive and objective evaluation of speech understanding: the "challenge" evaluation campaign of the I3 speech workgroup of the French CNRS", *Proceedings of the 3rd International Conference on Language Resources and Evaluation (LREC)*, ELDA, Las Palmas, Spain, pp. 529–535, May 2002.

[AON 00] AONE C., RAMOS-SANTACRUZ M., "REES: a large-scale relation and event extraction system", *First Annual Conference of the North American Chapter of the Association for Computational Linguistics*, San Francisco, CA, 2000.

[APP 93] APPELT D., HOBBS J., BEAR J. *et al.*, "FASTUS: a finite-state processor for information extraction from real-world text", *International Joint Conference on Artificial Intelligence IJCAI-93*, Chambéry, France, 1993.

[ARN 94] ARNOLD D., BALKAN L., MEIJER S. *et al.*, *Machine Translation: An Introductory Guide*, Blackwell, London, 1994.

[ASC 03] ASCHER D., LUTZ M., *Learning Python*, 2nd ed., O'Reilly, Sebastopol, 2003.

[ASL 82] ASLAM J., LEBLANC A., STEIN C., "Clustering data without prior knowledge", *4th International Workshop Algorithm Engineering*, Springer LNCS, 1982.

[ASS 93] ASSARAF A., "Quand dire, c'est lier", *Nouveaux Actes Sémiotiques, University of Limoges*, PULIM, no. 28, 1993.

[ATW 08] ATWELL E., Corpus linguistics and language learning: bootstrapping linguistic knowledge and resources from text, PhD Dissertation, The University of Leeds, 2008.

[AUG 22] AUGÉ C., *Petit Larousse Illustré, nouveau dictionnaire encyclopédique*, 185th ed., Larousse, Paris, 1922.

[AUS 55] AUSTIN J.L., *How to Do Things with Words*, Harvard University Press, Cambridge, MA (translated into French by Lane G., *Quand dire, c'est faire*, Le Seuil, Paris, 1970), 1955.

[BAC 89] BACH E., *Informal Lectures on Formal Semantics*, State University of New York, New York, 1989.

[BAC 00] BACHIMONT B., "Engagement sémantique et engagement ontologique : conception et réalisation d'ontologies en Ingénierie des connaissances", in CHARLET J., ZACKLAD M., KASSEL G. *et al.* (eds), *Ingénierie des connaissances, évolutions récentes et nouveaux défis*, Eyrolles, Paris, pp. 305–324, 2000.

[BAC 07] BACH N., BADASKAR S., "A review of relation extraction", Language Technologies Institute, Carnegie Mellon University, available at: http://orb.essex.ac.uk/CE/CE807/Readings/A-survey-on-Relation-Extraction.pdf, 2007.

[BAL 97] BALDWIN B., "CogNIAC: high precision coreference with limited knowledge and linguistic resources", *Proceedings of the ACL'97/EACL'97 Workshop on Operational Factors in Practical, Robust Anaphora Resolution*, Madrid, Spain, pp. 38–45, 1997.

[BAL 07] BALDRIDGE J., ASHER N., HUNTER J., "Annotation for and robust parsing of discourse structure on unrestricted texts", *Zeitschrift fur Sprachwissenschaft*, vol. 26, pp. 213–239, 2007.

[BAS 12] BASS L., CLEMENTS P., KAZMAN R., *Software Architecture in Practice*, Addison-Wesley, Upper Saddle River, 2012.

[BAT 00] BATLINER A., BUCOW J., NIEMANN H. *et al.*, "The prosody module", in WAHLSTER W. (ed.), *Verbmobil: Foundations of Speech-to-Speech Translation*, Springer, Berlin, 2000.

[BAY 00] BAYLON C., MIGNOT X., *Initiation à la sémantique du langage*, Nathan Université, Paris, 2000.

[BEL 05] BELL D., *Software Engineering for Students: A Programming Approach*, Addison-Wesley, Harlow, 2005.

[BEN 66] BENVENISTE E., *Problèmes de linguistique générale*, Gallimard, Paris, 1966.

[BER 95] BERRY M., DUMAIS S., O'BRIEN G., "Using linear algebra for intelligent information retrieval", *SIAM Review*, vol. 37, no. 4, pp. 573–595, 1995.

[BER 02] BERNSEN N.-O., BERTON A., CHARFUELÁN M. *et al.*, Progress report on the natural language understanding, Dialogue Management, Response Generation, and Speech Synthesis Components, Vico Deliverable D11, NISLab, November 2002.

[BER 04] BERRY M.J.A., LINOFF G.S., *Data Mining Techniques For Marketing, Sales, and Customer Relationship Management*, 2nd ed., Wiley Publishing, Indianapolis, 2004.

[BER 13] BERRENDONNER A., "Du morphème à la période: extension du domaine de la syntaxe", available at: cle.ens-lyon.fr, updated 9 December 2013.

[BIK 97] BIKEL D., MILLER S., SCHWARTZ R. *et al.*, "Nymble: a high-performance learning name-finder", *Proceedings of the Fifth Conference on Applied Natural Language Processing*, pp. 194–201, 1997.

[BIR 09] BIRD S., KLEIN E., LOPER E., *Natural Language Processing with Python: Analyzing Text with the Natural Language Toolkit*, O'Reilly Media, 2009.

[BLA 01] BLACKBURN P., DE RIJKE M., VENEMA Y., *Modal Logic*, Cambridge University Press, 2001.

[BLA 04] BLACKBURN P., BOS J., *Working with Discourse Representation Theory*, Center for the Study of Language and Information, 2004.

[BLA 05] BLACKBURN P., BOS J., *Representation and Inference for Natural Language: A First Course in Computational Semantics*, CSLI Press, Stanford, 2005.

[BLA 06] BLACK W., ELKATEB S., RODRIGUEZ H. *et al.*, "Introducing the Arabic WordNet project", *Third International WordNet Conference GWC-06*, Jeju Island, Korea, 22–26 January 2006.

[BOB 64] BOBROW D., "A question-answering system for high school algebra word problems", *AFIPS '64*, pp. 591–614, 27–29 October 1964.

[BOB 77] BOBROW D.G., WINOGRAD T., "An overview of KRL: a knowledge representation language", *Cognitive Science*, vol. 1, pp. 3–45, 1 January 1977.

[BOI 01] BOISSIER O., "Modèles et architectures d'agents", in BRIOT J.P., DEMAZEAU Y. (eds), *Principes et Architectures des Systèmes Multi-Agents*, Hermes-Lavoisier, 2001.

[BOI 02] BOITET C., MANGETO M., SÉRASSET G., "The Papillon project: cooperatively building lexical data-base to drive open source dictionaries & lexicon", *Proceedings of NLP and XML NLPXML 2002, COLING Workshop*, Taipei, Taiwan, pp. 9–15, 31 August 2002.

[BOL 95] BOLAND J.D., TANENHAUS M.K., GARNSEY S.M. *et al.*, "Verb argument structure in parsing and interpretation: evidence from wh-questions", *Journal of Memory and Language*, vol. 34, pp. 774–806, 1995.

[BOL 04] BOLSHAKOV I.A., GELBUKH A., *Computational Linguistics: Models, Resources, Applications*, Instituto Politécnico Nacional, Mexico, 2004.

[BOL 11] BOLLEGALA D., MATSUO Y., ISHIZUKA M., "Relation adaptation: learning to extract novel relations with minimum supervision", *Proceedings of the Twenty-Second International Joint Conference on Artificial Intelligence*, Barcelona, Spain, 16–22 July 2011.

[BOS 96] BOS J., GAMBACK B., LIESKE C. *et al.*, Compositional semantics in Verbmobil, Report 135 University of Sunderland Computational Linguistics, 1996.

[BOU 97] BOUILLON P., Polymorphie et sémantique lexical: le cas des adjectifs, PhD Thesis, Paris VII, Paris, 1997.

[BOU 99] BOUILLON P., "The adjective "vieux": the point of view of "generative lexicon"", in VIEGAS E. (ed.), *Breadth and Depth of Semantic Lexicons*, Kluwer Academic Press, 1999.

[BOU 02a] BOUFADEN N., LAPALME G., BENGIO Y., "Découpage thématique : un outil d'aide à l'extraction d'information", *TALN 2002*, Nancy, France, June 2002.

[BOU 02b] BOUFADEN N., LAPALME G., BENGIO Y., "Segmentation en thèmes de conversations téléphoniques : traitement en amont pour l'extraction d'information", *TALN02*, Nancy, 24–27 June 2002.

[BOY 77] BOYER R.S. *et al.*, "A fast string searching algorithm", *CACM*, vol. 20, no. 10, pp. 762–772, October 1977.

[BRA 79] BRACHMAN R.J., "On the epistemological status of semantic networks", in FINDLER N. (ed.), *Associative Networks: Representation and Use of Knowledge by Computers*, Academic Press, New York, 1979.

[BRA 85] BRACHMAN R.J., LEVESQUE H.J., *Readings in Knowledge Representation*, Morgan Kauffman, Los Altos, 1985.

[BRA 95] BRATT H., DOWDING J., HUNICKE-SMITH K., "The SRI telephone-based ATIS system", *Proceedings of the Spoken Language Systems Technology Workshop*, Austin, TX, January 1995.

[BRE 87] BRENNAN S.E., WALKER FRIEDMAN M.A., POLLARD C.J., "A centering approach to pronouns", *The 25th Annual Meeting of the Association for Computational Linguistics*, Stanford, CA, pp. 155–162, 1987.

[BRI 93] BRILL E., *A Corpus Based Approach to Language Learning*, PhD Dissertation, University of Pennsylvania, 1993.

[BRI 10] BRILLANT A., *XML : Cours et exercices*, Eyrolles, Paris, 2010.

[BRO 88] BROWN G., YULE G., *Discourse Analysis*, Cambridge University Press, Cambridge, 1988.

[BRO 91] BROWN P.F., LAI J.C., MERCER R.L., "Aligning sentences in parallel corpora", *Proceedings of ACL 92*, pp. 169–176, 1991.

[BUC 55] BUCHLER J. (ed.), *Philosophical Writings of Peirce*, The New Dover Edition, New York, 1955.

[BUN 07] BUNESCU R., MOONEY R.J., "Learning to extract relations from the web using minimal supervision", *Proceedings of the 45th Annual Meeting of the Association of Computational Linguistics*, pp. 576–583, 2007.

[BUR 04] BURCHARDT A., KOLLER A., WALTER S., *Computational Semantics*, ESSLLI04, Nancy, 2004.

[BUR 15] BURNARD L., BAUMAN S. (eds), "TEI P5: guidelines for electronic text encoding and interchange by the TEI consortium", *TEI Consortium*, available at: http://www.tei-c.org/release/doc/tei-p5-doc/readme-2.8.0.html, 2015.

[BUS 96] BUSA F., Compositionality and the semantics of nominals, PhD Dissertation, Brandeis University, 1996.

[CAE 79] CAELEN J., Un modèle d'oreille : analyse de la parole, reconnaissance phonémique, PhD thesis, Toulouse, 1979.

[CAE 94] CAELEN J., CAILLAUD B., ANTOINE J.-Y., "Projet Micro: Modélisation informatique de la cognition en reconnaissance de l'oral", *Actes du séminaire Reconnaissance, GDR-PRC Communication Homme-machine*, Nancy, pp. 295–319, 1994.

[CAE 95] CAELEN J., "Vers une logique dialogique", *Séminaire International de Pragmatique*, Jerusalem, 1995.

[CAM 12] CAMBRIA E., LIVINGSTONE A., HUSSAIN A., "The hourglass of emotions", in ESPOSITO A. *et al.* (eds), *Cognitive Behavioural Systems 2011*, LNCS 7403, Springer-Verlag, Berlin, Heidelberg, pp. 144–157, 2012.

[CAM 15] CAMBRIA E., HUSSAIN A., *Sentic Computing: A Common-Sense-Based Framework for Concept-Level Sentiment Analysis*, Springer, 2015.

[CAN 10] CANDITO M.-H., CRABBÉ B., DENIS P., "Statistical French dependency parsing: treebank conversion and first results", *Proceedings of LREC'10 Conference*, La Valletta, Malta, 2010.

[CAR 88] CARPENTER P.A., JUST M.A., "The role of working memory in language comprehension", in KLAHR D., KOTOVSKY K. (eds), *Complex Information Processing: the Impact of Herbert A. Simon*, Erlbaum, Hillsdale, 1988.

[CAR 03a] CARL M., WAY A. (eds), *Recent Advances in Example-Based Machine Translation*, vol. 21, Kluwer Academic Publishers, Dordrecht, 2003.

[CAR 03b] CARLSON L., MARCU D., OKUROWSKI M.E., "Building a discourse-tagged corpus in the framework of rhetorical structure theory", in VAN KUPPEVELT J., SMITH R. (eds), *Current Directions in Discourse and Dialogue*, Kluwer, New York, 2003.

[CAR 05] CARDEÑOSA J., GELBUKH A., TOVAR E. (eds), *Universal Networking Language: Advances in Theory and Applications*, Instituto Politécnico Nacional, Mexico, 2005.

[CAV 94] CAVNAR W.B., TRENKLE J.M., "N-gram-based text categorization", *Proceedings of SDAIR-94, 3rd Annual Symposium on Document Analysis and Information Retrieval*, 1994.

[CAV 98] CAVAZZA M., "Synchronous TFG for speech translation", *Proceedings of the International TAG Workshop*, Philadelphia, PA, 28–31 July 1998.

[CHA 76] CHAFE W.L., "Givenness, contrastiveness, definiteness, subjects, topics and point of view", in CHARLES N.L. (ed.), *Subject and Topic*, Academic Press, New York, 1976.

[CHA 78] CHAROLLES M., "Introduction aux problèmes de la cohérence des textes", *Langue française*, vol. 38, pp. 7–41, 1978.

[CHA 86] CHAROLLES M., "Le problème de la cohérence dans les études sur le discours", in CHAROLLES M., PETOFI J.S., SOZER E. (eds), *Research in Text Connexity and Text Coherence*, Buske, Hamburg, pp. 1–65, 1986.

[CHA 87] CHAFE W.L., "Cognitive constraints on information flow", in RUSSELL S.T. (ed.), *Coherence and Grounding in Discourse*, John Benjamins, Amsterdam, 1987.

[CHA 08] CHAMBERS N., JURAFSKY D., "Unsupervised learning of narrative event chains", *Annual Meeting of the Association for Computational Linguistics: Human Language Technologies*, Columbus, OH, 2008.

[CHE 93] CHEN S.F., "Aligning sentences in bilingual corpora using lexical information", *ACL93*, Columbus, OH, 22–26 June 1993.

[CHI 92] CHINCHOR N., "MUC 4 evaluation metrics", *The Fourth Message Understanding Conference (MUC-4)*, pp. 22–29, 1992.

[CHI 02] CHICOISNE G., Dialogue entre agents naturels et agents artificiels, PhD Thesis, Institut National Polytechnique de Grenoble, 2002.

[CHO 57] CHOMSKY N., *Syntactic Structures*, Mouton The Hague, Paris, 1957.

[CHO 65] CHOMSKY N., *Aspects of the Theory of Syntax*, MIT Press, Cambridge, 1965.

[CHO 71] CHOMSKY N., "Deep structure, surface structures, and semantic interpretation", in STEINBERG D., JACOBOVITS L.A. (eds), *Semantics*, Cambridge University Press, Cambridge, 1971.

[CHR 84] CHRISTODOULAKIS S., FALOUTSOS C., "Design considerations for a message file server", *IEEE Transactions on Software Engineering*, vol. SE-10, no. 2, pp. 201–210, 1984.

[CHU 40] CHURCH A., "A formulation of the simple theory of types", *Journal of Symbolic Logic*, vol. 5, pp. 56–68, 1940.

[CHU 06] CHUN H.-W., TSURUOKA Y., KIM J.-D. *et al.*, "Extraction of gene-disease relations from Medline using domain dictionaries and machine learning", *Pacific Symposium on Biocomputing*, pp. 4–15, 2006.

[CIA 01] CIANCARINI P., WOOLDRIDGE M. (eds), *Agent-Oriented Software Engineering*, Springer-Verlag, January 2001.

[CLA 79] CLARK H.H., SENGAL C.J., "In search of referents for nouns and pronouns", *Memory and Cognition*, vol. 7, pp. 35–41, 1979.

[CLE 16a] CLERC M., BOUGRAIN L., LOTTE F., *Brain-Computer Interfaces 1: Foundations and Methods*, ISTE Ltd, London and John Wiley & Sons, New York, 2016.

[CLE 16b] CLERC M., BOUGRAIN L., LOTTE F., *Brain-Computer Interfaces 2: Technology and Applications*, ISTE Ltd, London and John Wiley & Sons, New York, 2016.

[COL 97] COLINEAU N., Étude des marqueurs discursifs dans le dialogue finalisé, Thesis, Joseph Fourier University, 1997.

[COO 71] COOK W.A., A case grammar matrix, Languages and Linguistics: Working Papers, no. 5, Washington University School of Languages and Linguistics, pp. 50–81, 1971.

[COO 89] COOK W.A., *Case Grammar Theory*, Georgetown University Press, Washington, D.C., 1989.

[COP 92] COPESTAKE A., "The ACQUILEX LKB: representation issues in the semi-automatic acquisition of large lexicons", *3rd Conference on Applied Natural Language Processing*, Trento, Italy, 1992.

[COR 95] CORTES C., VAPNIK V., "Support-vector networks", *Machine Learning*, vol. 20, pp. 273–297, 1995.

[COV 94] COVINGTON M., *Natural Language Processing for Prolog Programmers*, Prentice Hall, Englewood, 1994.

[CRA 85] CRAIN S., STEEDMAN M., "On not being led up by the garden path: the use of context by the psychological syntax processor", in DOWTY D.R. *et al.* (eds), *Natural Language Parsing: Computational and Theoretical Perspectives*, Cambridge University Press, Cambridge, 1985.

[CRA 90] CRAWLEY R.A., STEVENSON R.J., KLEINMAN D., "The use of heuristic strategies in the interpretation of pronouns", *Journal of Psycholinguistic Research*, vol. 19, pp 245–264, 1990.

[CRI 98] CRISTEA D., IDE N., ROMARY L., Veins theory: a model of global discourse cohesion and coherence, *Proceedings of COLING 1998*, Montreal, Canada, 1998.

[CRO 93] CROFT W., *Typology and Universals*, Cambridge University Press, Cambridge, 1993.

[CRO 96] CROCKER M., Mechanisms for sentence processing, Research Paper EUCCS/RP-70, Centre for Cognitive Science, University of Edinburgh, 1996.

[CRU 00a] CRUSE A., "Lexical "facets": between monosemy and polysemy", in BECKMANN S., KONIG P.P., WOLF T. (eds), *Sprachspiel und Bedeutung: Festschrift für Franz Hundsnurscher zum 50 Geburtstag*, Max Niemeyer Verlag, Tübingen, pp. 25–36, 2000.

[CRU 00b] CRUSE A., *Meaning in Language: an Introduction to Semantics and Pragmatics,* Oxford University Press, 2000.

[CRU 04] CRUSE A., "Lexical facets and metonymy", *Ilha do desteros Journal of English Language, Literatures in English and Cultural Studies*, no. 47, pp. 073–096, 2004.

[CRY 91] CRYSTAL D. *A Dictionary of Linguistics and Phonetics*, 3rd ed., Blackwell, London, 1991.

[CUL 04] CULOTTA A., SORENSEN J., "Dependency tree kernels for relation extraction", *Proceedings of the 42nd Meeting of the Association for Computational Linguistics (ACL'04)*, pp. 423–429, 2004.

[CUM 07] CUMMINS R., O'RIORDAN C., "An axiomatic study of learned termweighting schemes", *Artificial Intelligence Review*, vol. 28, no. 1, pp. 51–68, 2007.

[DAE 05] DAELEMANS W., VAN DEN BOSCH A., *Memory-Based Language Processing*, Cambridge University Press, Cambridge, 2005.

[DAW 03] DAWSON M., *Python Programming for the Absolute Beginner*, Premier Press, Boston, 2003.

[DE 16] DE SAUSSURE F. *Cours de linguistique générale*, Payot & Rivages, Paris, 1916.

[DE 81] DE BEAUGRANDE R., DRESSIER W.V., *Introduction to Text Linguistics*, Longman, 1981.

[DEK 12] DEKKER P., *Dynamic Semantics*, Springer, Dordrecht, 2012.

[DEO 04] DEOSKAR T., "Techniques for anaphora resolution: a survey", Technical Report CS, Cornell University, available at: www.cs.cornell.edu/courses/cs674/2005sp/projects/tejaswini-deoskar.doc, 17 May 2004.

[DER 16] DERCZYNSKI L., STRÖTGEN J., MAYNARD D. *et al.*, "GATE-time: extraction of temporal expressions and events", *Proceedings of LREC*, 2016.

[DEW 98] DEWE J., KARLGREN J., BRETAN I., "Telia research, assembling a balanced corpus from the Internet", available at: http://eprints.sics.se/63/1/Dropjaw_korpus.html, 1998.

[DIE 97] DIETTERICH T.G., LATHROP R.H., LOZANO-PEREZ T., "Solving the multiple instance problem with axis-parallel rectangles", *Artificial Intelligence*, vol. 89, nos. 1–2, pp. 31–71, 1997.

[DME 07] D'MELLO S., PICARD R.W., GRAESSER A., "Toward an affect-sensitive AutoTutor", *IEEE Intelligent Systems*, vol. 22, no. 4, pp. 53–61, 2007.

[DOS 55] DOSTERT L.E., "The Georgetown–I.B.M. experiment", in LOCKE W.N., BOOTH A.D. (eds), *Machine Translation of Languages*, Wiley, 1955.

[DOY 85] DOYON G., TALBOT P., *La logique du raisonnement*, Le Griffon d'Argile, Sainte-Foy, 1985.

[DUB 71] DUBOIS J. *et al.*, *Dictionnaire du français contemporain*, Larousse, Paris, 1971.

[DUB 91] DUBOIS D. (ed.), *Sémantique et cognition : catégories, prototypes, typicalité*, Editions du CNRS, Paris, 1991.

[DUC 69] DUCROT O., "Présupposés et sous-entendus", *Langue française*, no. 4, pp. 30–43, 1969.

[DUC 87] DUCROT O., *Le Dire et le Dit*, Editions de Minuit, Paris, 1987.

[DUT 04] DUTEIL-MOUGEL C., "Introduction à la sémantique interprétative", *Texto !* [e-book], available at: http://www.revue-texto.net/Reperes/Themes/Duteil/Duteil_Intro.html, December 2004.

[DUT 08] DUTTA K. *et al.*, "Resolving pronominal anaphora in Hindi using Hobbs' algorithm", *Web Journal of Formal Computation and Cognitive Linguistics*, no. 10, 2008.

[EBR 16] EBRAHIMI J., DOU D., LOWD D., "A joint sentiment-target-stance model for stance classification in tweets", *Proceedings of COLING 2016, the 26th International Conference on Computational Linguistics*: Technical Papers, pp. 2656–2665, Osaka, Japan, available at: https://www.aclweb.org/anthology/C/C16/C16-1250.pdf, 11–17 December 2016.

[EDW 93] EDWARDS J., "Principles and contrasting systems of discourse transcription", in EDWARDS J., LAMPERT M. (eds), *Talking Data: Transcription and Coding in Discourse Research*, Lawrence Erlbaum Associates, Hillsdale, 1993.

[EIB 16] EIBE F., HALL M.A., WITTEN I.H., *The WEKA Workbench. Online Appendix for Data Mining: Practical Machine Learning Tools and Techniques*, Morgan Kaufmann, 4th ed., 2016.

[EIS 08] EISENSTEIN J., BARZILAY R., "Bayesian unsupervised topic segmentation", *Proceedings of the Conference on Empirical Methods in Natural Language Processing (EMNLP '08)*, Honolulu, HI, pp. 334–343, 2008.

[EKL 96] EKLUND P.W., ELLIS G., MANN G. (eds), *Conceptual Structures: Knowledge Representation as Interlingua*, Springer-Verlag, Berlin, 1996.

[EKM 92] EKMAN P., "An argument for basic emotions", *Cognition Emotion*, vol. 6, no. 3, pp. 169–200, 1992.

[ELL 92] ELLIOTT C., The affective reasoner: a process model of emotions in a multi-agent system, PhD Thesis, Northwestern University, The Institute for the Learning Sciences, Technical Report No. 32, May 1992.

[ENC 09] ENCARTA DVD, *Dictionnaire français*, Microsoft Corporation, 2009.

[ERM 80] ERMAN L.D., HAYES-ROTH F., LESSER V.R. *et al.*, "The Hearsay-II speech understanding system: integrating knowledge to resolve uncertainty", *ACM Computing Surveys*, vol. 12, no. 2, pp. 213–251, 1980.

[ERM 81] ERMAN L.D., LONDON P.E., FICKAS S.F., "The design and an example use of Hearsay-II", *Proc. IJCAI-81*, pp. 409–415, 1981.

[ESU 06] ESULI A., SEBASTIANI F., "SENTIWORDNET: a publicly available lexical resource for opinion mining", *The 5th Conference on Language Resources and Evaluation LREC*, Genoa, Italy, 24–26 May 2006.

[ETZ 05] ETZIONI O., CAFARELLA M., DOWNEY D., "Unsupervised named-entity extraction from the web: an experimental study", *Artificial Intelligence*, vol. 165, no. 1, pp. 91–134, 2005.

[FAU 84] FAUCONNIER G., *Espaces Mentaux*, Editions de Minuit, Paris, 1984.

[FEL 90a] FELLBAUM C., MILLER G., "Folk psychology or semantic entailment? A reply to Rips and Conrad", *Psychological Review*, vol. 97, no. 4, pp. 565–570, 1990.

[FEL 90b] FELLBAUM C., "English verbs as a semantic net", *International Journal of Lexicography*, vol. 3, no. 4, pp. 278–301, 1990.

[FEL 05] FELLBAUM C., "WordNet and wordnets", in BROWN K. *et al.* (eds), *Encyclopedia of Language and Linguistics*, 2nd ed., Elsevier, Oxford, 2005.

[FEL 07] FELDT K., *Programming Firefox: Building Rich Internet Applications with XUL*, O'Reilly, 2007.

[FER 00] FERNANDEZ A.J., HILL P.M., "A comparative study of eight constraint programming languages over the Boolean and finite domains", *Constraints*, vol. 5, no. 3, pp. 275–301, 2000.

[FIL 66] FILLMORE C.J., "A proposal concerning English prepositions", *Monograph Series on Languages and Linguistics*, Georgetown University, no. 19, pp. 19–33, 1966.

[FIL 68] FILLMORE C.J., "The case for case", in BACH E., HARMS R.T. (eds), *Universals in Linguistic Theory*, Holt, Rinehart, and Winston, New York, 1968.

[FIL 71] FILLMORE C.J., "Types of lexical information", in STEINBERG D., JOCABOVITS L. (eds), *Semantics*, Cambridge University Press, Cambridge, 1971.

[FIL 77] FILLMORE C.J., "The case for case reopened", in COLE P., SADOCK J. (eds), *Syntax and Semantics*, Academic Press, New York, vol. 8, 1977.

[FIN 09] FINLAYSON M., "Deriving narrative morphologies via analogical story merging", *Proceedings of the 2nd International Conference on Analogy*, Sofia, Bulgaria, 2009.

[FLI 00] FLICKINGER D., COPESTAKE A., SAG I.A., "HPSG analysis of English", in WAHLSTER W. (ed.), *Verbmobil: Foundations of Speech-to-Speech Translation*, Springer, Berlin, 2000.

[FOD 64] FODOR J., KATZ J., *The Structure of Language*, Prentice-Hall, New Jersey, 1964.

[FOL 96] FOLTZ P.W., "Latent semantic analysis for text-based research", *Behavior Research Methods, Instruments, and Computers*, vol. 28, no. 2, pp. 197–202, 1996.

[FOR 65] FORGY E.W., "Cluster analysis of multivariate data: efficiency versus interpretability of classifications", *Biometrics*, vol. 21, pp. 768–769, 1965.

[FOR 79] FORSTER K.I., "Levels of processing and the structure of the languages processor", in COOPER W.E., WALKER E.C.T. (eds), *Sentence Processing: Psycholinguistic Studies*, Lawrence Erlbaum Associates, 1979.

[FRA 92] FRAKES W.B., BAEZA-YATES R. (eds), *Information Retrieval: Data Structures & Algorithms*, Prentice Hall, Upper Saddle River, 1992.

[FRA 06] FRANCOPOULO G., MONTE G., CALZOLARI N. *et al.*, "Lexical markup framework (LMF)", *Proceedings of the 4th International Conference on Language Resources and Evaluation (LREC)*, Genoa, Italy, 24–26 May 2006.

[FRA 08] FRANK R., MATHIS D., BADECKER W., The acquisition of anaphora by simple recurrent networks, Manuscript, Département de linguistique, Yale University, available at: http://whitney.ling.yale.edu/~rfrank/Bob_Frank/Publications.html, 2008.

[FRI 96] FRIEDMAN N., GOLDSZMIDT M., "Building classifiers using Bayesian networks", *Proceedings of the National Conference on Artificial Intelligence*, AAAI Press, Menlo Park, CA, pp. 1277–1284, 1996.

[FUC 91] FUCHS C., "Polysémie, interprétation et typicalité : l'exemple de pouvoir", in DUBOIS D. (ed.), *Sémantique et cognition : catégories, prototypes, typicalité*, éditions du CNF, Paris, 1991.

[FUC 96] FUCHS C., *Les ambigüités du français*, Ophrys, Paris 1996.

[FUC 07] FUCHS C., "Champ sémantique et champ lexical", *Encyclopædia Universalis*, 2007.

[FUN 94] FUNG P., MCKEOWN K., "Aligning noisy parallel corpora across language groups: word pair feature matching by dynamic time warping", *Proceedings of the Association for Machine Translation (AMTA) in the Americas*, pp. 81–88, 1994.

[GAL 93] GALE W.A., CHURCH K.W., "A program for aligning sentences in bilingual corpora", *Computational Linguistics*, vol. 19, pp. 75–102, 1993.

[GAL 03] GALLEY M., MCKEOWN K.R., FOSLER-LUSSIER E. *et al.*, "Discourse segmentation of multi-party conversation", *Proceedings of ACL*, pp. 562–569, 2003.

[GAN 03] GANGEMI A., GUARINO N., "Sweetening WORDNET with DOLCE", *AI Magazine*, vol. 24, no. 3, pp. 13–24, 2003.

[GAR 93] GARLAN D., SHAW M., "An introduction to software architecture", in AMBRIOLA V., TORTORA G. (eds), *Advances in Software Engineering and Knowledge Engineering*, vol. 1, World Scientific Publishing Company, New Jersey, 1993.

[GAR 03] GARRIDO J., "Relevance versus connection: discourse and text as units of analysis", *CÍRCULO clac de lingüística aplicada a la comunicación*, vol. 13, 2003.

[GAV 00a] GAVALDA M., "SOUP: a parser for real-world spontaneous speech", *Proceedings of the Sixth International Workshop on Parsing Technologies (IWPT-2000)*, Trento, Italy, February 2000.

[GAV 00b] GAVALDA M., Growing semantic grammars, PhD Dissertation, Carnegie Mellon University, 2000.

[GAZ 89a] GAZDAR G., *Natural Language Processing in Lisp: An Introduction to Computational Linguistics*, Addison-Wesley Longman Publishing Co., Inc., Boston, MA, 1989.

[GAZ 89b] GAZDAR G., *Natural Language Processing in Prolog: An Introduction to Computational Linguistics*, Addison-Wesley Longman Publishing Co., 1989.

[GEN 07] GENETTE G., *Discours du récit*, Le Seuil, Paris, 2007.

[GEU 02] GEUTNER P., STEFFENS F., MANSTETTEN D., "Design of the Vico spoken dialog system: evaluation of user expectations by Wizard of Oz simulations", *Proceedings of LREC02 Conference*, Las Palmas, Spain, 2002.

[GÓM 95] GÓMEZ-PÉREZ A., JURISTO N., PAZOS J., "Evaluation and assessment of knowledge sharing technology", in MARS N.J. (ed.), *Towards Very Large Knowledge Bases, Knowledge Building and Knowledge Sharing*, IOS Press, Amsterdam, 1995.

[GOR 93] GORDON P.C., GROSZ B.J., GILLIOM L.A., "Pronouns, names and the centering of attention in discourse", *Cognitive Science*, vol. 17, no. 3, pp. 311–347, 1993.

[GOR 06] GORTON I., *Essential Software Architecture*, Springer, Berlin, 2006.

[GOR 09] GORDON C., *Making Meanings, Creating Family: Intertextuality and Framing in Family Interaction*, Oxford University Press, Oxford, 2009.

[GRA 05] GRAESSER A.C., CHIPMAN P., HAYNES B.C. *et al.*, "AutoTutor: an intelligent tutoring", *IEEE Transactions on Education*, vol. 48, no. 4, pp. 612–618, 2005.

[GRA 10] GRANDY R., OSHERSON D., "Sentential logic primer", Online Technical Report, University of Princeton, 2004, available at: http://www.princeton.edu/~osherson/primer.pdf, 2010.

[GRE 66] GREIMAS A., *Sémantique structurale*, PUF, Paris 1966.

[GRE 68] GREIMAS A., RASTIER F., "The interaction of semiotic constraints", *Yale French Studies*, vol. 41, pp. 86–105, 1968.

[GRE 04a] GRENON P., SMITH B., GOLDBERG L., "Biodynamic ontology: applying BFO in the biomedical domain", in PISANELLI D.M. (ed.), *Ontologies in Medicine*, IOS Press, Amsterdam, 2004.

[GRE 04b] GRENON P., SMITH B., "SNAP and SPAN: towards dynamic spatial ontology", *Spatial Cognition and Computation*, vol. 4, pp. 69–103, 2004.

[GRE 06] GREGORY CAPORASO J., BAUMGARTNER W.A., KIM JR. H., "Concept recognition, information retrieval, and machine learning in genomics question-answering", *TREC06 Proceedings*, 2006.

[GRI 06] GRIES S.T., STEFANOWITSCH A. (eds), *Corpora in Cognitive Linguistics: Corpus-Based Approaches to Syntax and Lexis*, Mouton de Gruyter, 2006.

[GRI 07] GRIMM S., HITZLER P., ABECKER A., "Knowledge representation and ontologies: logic, ontologies and semantic web languages, semantic web services", in STUDER R., GRIMM S., ABECKER A. (eds), *Knowledge Representation and Ontologies*, Springer, Berlin, Heidelberg, 2007.

[GRI 10] GRISHMAN R., "Information extraction", in CLARK A., FOX C., LAPPIN S. (eds), *The Handbook of Computational Linguistics and Natural Language Processing*, Wiley-Blackwell, Malden, 2010.

[GRO 77] GROSZ B.J., The representation and use of focus in dialogue understanding, Technical Report 151, SRI International, 333 Ravenswood Ave, Menlo Park, CA 94025, 1977.

[GRO 83] GROSZ B.J., JOSHI ARAVIND K., WEINSTEIN S., "Providing a unified account of definite noun phrases in discourse", *Proceedings, 21st Annual Meeting of the Association of Computational Linguistics*, pp. 44–50, 1983.

[GRO 86] GROSZ B.J., SIDNER C.L., "Attentions, intentions and the structure of discourse", *Computational Linguistics*, vol. 12, pp. 175–204, 1986.

[GRO 91] GROENENDIJK J., STOKHOF M., "Dynamic predicate logic", *Linguistics and Philosophy*, vol. 14, pp. 39–100, 1991.

[GRO 04] GROSSMAN D.A., FRIEDER O., *Information Retrieval, Algorithms and Heuristics*, Springer, 2004.

[GRU 95] GRUBER T.R., "Towards principles for the design of ontologies used for knowledge sharing", *International Journal of Human Computer Studies*, vol. 43, nos. 5/6, pp. 907–928, 1995.

[GRÜ 95] GRÜNINGER M., FOX M.S., "Methodology for the design and evaluation of ontologies", *IJCAI-95 Workshop on Basic Ontological Issues in Knowledge Sharing*, Montreal, 19–20 August 1995.

[GUA 95] GUARINO N., GIARETTA P., "Ontologies and knowledge bases: towards a terminological clarification", in MARS N.J.I. (ed.), *Towards Very Large Knowledge Bases*, IOS Press, Amsterdam, 1995.

[GUA 98] GUARINO N., "Formal ontology in information systems", *Proceedings of FOIS'98*, Trento, Italy, pp. 3–15, 6–8 June 1998.

[GUP 12] GUPTA P., GUPTA V., "A survey of text question answering techniques", *International Journal of Computer Applications (0975 – 8887)*, vol. 53, no. 4, September 2012.

[HAB 97] HABERT B., NAZARENKO A., SALEM A., *Les linguistiques de corpus*, Armand Colin, Paris, 1997.

[HAJ 98] HAJIC J., "Building a syntactically annotated corpus: the Prague dependency treebank", in HAJICOVA E. (ed.), *Issues of Valency and Meaning, Studies in Honor of Jarmila Panevova*, Charles University Press, Prague Karolinum, 1998.

[HAL 64] HALLIDAY M.A.K., "The linguistic study of literary texts", in LUNT H.G. (ed.), *Proceedings of the Ninth International Congress of Linguists*, Mouton, The Hague, pp. 302–307, 1964.

[HAL 70] HALLIDAY M.A.K., "Language structure and language function", in LYONS J. (ed.), *New Horizons in Linguistics*, Penguin Books, Harmondsworth, 1970.

[HAL 94] HALLIDAY M.A.K., *An Introduction to Functional Grammar*, Edward Arnold, London, 1994.

[HAR 51] HARRIS J., *Hermes: or a Philosophical Inquiry concerning Language and Universal Grammar*, Scolar Press, Menston, 1968.

[HAR 52] HARRIS Z., "Discourse analysis", *Language*, vol. 28, pp. 18–23, 1952.

[HAR 97] HARDT D., "An empirical approach to VP ellipsis", *Computational Linguistics*, vol. 23, no. 4, 1997.

[HAS 68] HASAN R., *Grammatical Cohesion in Spoken and Written English, Part 1, Program in Linguistics, and English Teaching*, Paper No. 7, Longman, London, 1968.

[HEA 92] HEARST M.A., "Automatic acquisition of hyponyms from large text corpora", *Proceedings of the 14th International Conference on Computational Linguistics*, Nantes, France, pp. 539–545, 1992.

[HEA 97] HEARST M., "TextTiling: segmenting text into multi-paragraph subtopic passages", *Computational Linguistics*, vol. 23, no. 1, pp. 33–64, 1997.

[HÉB 02] HÉBERT L., "La Sémantique interprétative en résumé", *Texto !* [e-book], Disponible sur, available at : http://www.revue-texto.net/Reperes/Themes/Hebert_SI.html, June 2002.

[HEB 12] HEBERT L., "Dictionnaire de sémiotique générale", *texto! Textes & Cultures* [en ligne], vol. XVII, nos. 1 and 2, available at: http://www.signosemio.com/documents/dictionnaire-semiotique-generale.pdf, 2012.

[HEI 82] HEIM I., The semantics of definite and indefinite noun phrases, PhD Thesis, University of Massachusetts, Amherst, 1982.

[HIN 00] HINRICHS E.W., KÜBLER S., KORDONI V. *et al.*, "Robust chunk parsing for spontaneous speech", in WAHLSTER W. (ed.), *Verbmobil: Foundations of Speech-to-Speech Translation*, Springer, Berlin, 2000.

[HIR 96] HIRSCHBERG J., NAKATANI C.H., "A prosodic analysis of discourse segments in direction-giving monologues", *The 34th Annual Meeting on Association for Computational Linguistics ACL96*, pp. 286–293, 1996.

[HIR 98] HIRSCHMAN L., "Language understanding evaluations: lessons learned from MUC and ATIS", *Proceedings of the 1st International Conference on Language Resources and Evaluation (LREC)*, Granada, Spain, pp. 117–122, 1998.

[HJE 43] HJELMSLEV L., Omkring sprogteoriens grundlæggelse, Festskrift udgivet af Københavns Universitet i anledning af Universitetets Aarsfest, University of Copenhagen, Copenhagen, 1943.

[HOB 76] HOBBS J.R., Pronoun resolution, Research Report 76-1, Department of Computer Sciences, City College, City University of New York, August 1976

[HOB 78] HOBBS J.R., "Resolving pronoun references", *Lingua*, vol. 44, pp. 311–338, 1978.

[HOL 97] HOLLARD S., L'organisation des connaissances dans le dialogue orienté par la tâche, Technical report 1-97, Geod CLIPS-IMAG, Grenoble, 1997.

[HOR 07] HORAK A., RAMBOUSEK A., "Dictionary management system for the DEB development platform", *Proceedings of the 4th International Workshop on Natural Language Processing and Cognitive Science*, pp. 129–138, 2007.

[HSI 95] HSINCHUN C., "Machine learning for information retrieval: neural networks, symbolic learning, and genetic algorithms", *Journal of the American Society for Information Science*, vol. 46, no. 3, pp. 194–216, 1995.

[HUA 13] HUANG Y., "Bayesian probabilistic model of discourse anaphoric comprehension linguistic typology, and neo-Gricean pragmatics", *Theoretical Linguistics*, vol. 39, nos. 1–2, 2013.

[HUD 88] HUDSON-D'ZMURA S.B., The structure of discourse and anaphor resolution: The discourse center and the roles of nouns and pronouns, PhD Thesis, University of Rochester, 1988.

[HUR 13] HURWITZ J., NUGENT A., HALPER F. *et al.*, *Big Data for Dummies*, John Wiley & Sons, Inc., Hoboken, NJ, 2013.

[HUT 92] HUTCHINS J.W., SOMERS H.L., *An Introduction to Machine Translation*, Academic Press, London, 1992.

[HUT 04] HUTCHINS J.W., "Machine translation and computer-based translation tools: what's available and how it's used", edited transcript of a presentation at the University of Valladolid (Spain), available at: http://www.hutchinsweb.me.uk/Valladolid-2004.pdf, March 2003.

[IDE 95] IDE N., VERONIS J., *Text Encoding Initiative: Background and Context*, Kluwer Academic Publishers, Dordrecht, 1995.

[ISO 86] ISO, ISO 8879 Standard, available at: http://www.iso.org/iso/fr/iso_catalogue/catalogue_tc/catalogue_detail.htm?csnumber=16387, 1986.

[JAC 72] JACKENDOFF R., *Semantic Interpretation in Generative Grammar*, MIT Press, Cambridge, 1972.

[JAM 84] JAMES W., "What is an emotion?", *Mind*, vol. 9, pp. 188–205, available at: http://psychclassics.yorku.ca/James/emotion.htm, 1884.

[JAN 16] JANSSOONE T., CLAVEL C., BAILLY K. *et al.*, "Using temporal association rules for the synthesis of embodied conversational agents with a specific stance", *International Conference on Intelligent Virtual Agents*, Springer International Publishing, September 2016.

[JOH 95] JOHNSON E., "The text encoding initiative", *TEXT Technology*, vol. 5, no. 3, pp. 174–175, Autumn 1995.

[JON 98] JONES D., BENCH-CAPON T., VISSER P., "Methodologies for ontology development", *15th FIP World Computer Congress*, pp. 62–75, 1998.

[JOS 81] JOSHI A.K., WEINSTEIN S., "Control of inference: role of some aspects of discourse structure-centering", *International Joint Conference on Artificial Intelligence IJCAI*, Vancouver, Canada, 1981.

[KAM 75] KAMP J.A.W., "Two theories about adjectives", in KEENAN E. (ed.), *Formal Semantics of Natural Language*, Cambridge University Press, Cambridge, 1975.

[KAM 81] KAMP H., "A theory of truth and semantic representation", in GROENENDIJK J.A.G., JANSSEN T.M.V., STCKHOF M.B.J. (eds), *Formal Methods in the Study of Language*, Mathematical Centre Tracts 135, Amsterdam, 1981.

[KAM 93] KAMP H., REYLE U., *From Discourse to Logic*, Kluwer, Dordrecht, 1993.

[KAM 04a] KAMBHATLA N., "Combining lexical, syntactic, and semantic features with maximum entropy models for extracting relations", *Proceedings of the 42nd Annual Meeting of the ACL*, Barcelona, Spain, 21–26 July 2004.

[KAM 04b] KAMPS J., MARX M., MOKKEN R.J. *et al.*, "Using wordnet to measure semantic orientation of adjectives", *The 4th International Conference on Language Resources and Evaluation*, LREC, Lisbon, Portugal, 26–28 May 2004.

[KAP 00] KAPLAN F., L'émergence d'un lexique dans une population d'agents autonomes, PhD Thesis, University of Paris 6, 2000.

[KAR 00] KARGER R., WAHLSTER W. (eds), "Facts and figures about the Verbmobil project, in Wolfgang Wahlster", *Verbmobil: Foundations of Speech-to-Speech Translation*, Springer, Berlin, 2000.

[KAT 64] KATZ J., POSTAL P., *An Integrated Theory of Linguistic Descriptions*, MIT Press, Cambridge, 1964.

[KEE 71] KEENAN E.L., "Two kinds of presupposition in natural language", in FILLMORE C.J., LANGENDOEN D.T. (eds), *Langendoen Studies in Linguistic Semantics*, Holt, Rinehart & Winston, New York, 1971.

[KEE 05] KEENAN E., "How much logic is built in natural language", *Fifteenth Amsterdam Colloquium, ILLC*, University of Amsterdam, pp. 39–45, 2005.

[KEM 75] KEMPSON R., *Presupposition and the Delimitation of Semantics*, Cambridge University Press, Cambridge, 1975.

[KEN 13] KENDALL K.E., KENDALL J.E., *Systems Analysis and Design*, 9th ed., Pearson, 2013.

[KER 96] KERBRAT-ORECCHIONI C., *La conversation*, Le Seuil, Paris, 1996.

[KIM 03] KIM J.-D., OHTA T., TATEISI Y. *et al.*, "GENIA corpus: a semantically annotated corpus for bio-textmining", *Eleventh International Conference on Intelligent Systems for Molecular Biology*, Brisbane, Australia, 29 June–3 July 2003.

[KIM 04] KIM S.-M., HOVY E., "Determining the sentiment of opinions", *The 20th International Conference on Computational Linguistics COLING*, Geneva, 23–27 August 2004.

[KIN 78] KINTSCH W., VAN DIJK T., "Towards a model of text comprehension and production", *Psychological Review*, vol. 85, pp. 363–394, 1978.

[KIN 98] KING M., MAEGAARD B., "Issues in natural language systems evaluation", *First International Conference on Language Resources and Evaluation (LREC)*, Granada Spain, vol. 1, pp. 225–230, 1998.

[KIP 05] KIPPER K., VerbNet: a broad-coverage, comprehensive verb lexicon, PhD Dissertation University of Pennsylvania, 2005.

[KLE 90] KLEIBER G., *La sémantique du prototype : catégories et sens lexical*, PUF, Paris, 1990.

[KLE 91] KLEIBER G., "Anaphore – deixis : où en sommes-nous?", *L'information grammaticale*, vol. 51, October 1991.

[KLE 96] KLEIBER G., "Cognition, sémantique et facettes: une "histoire" de livres et de romans", in KLEIBER G., RIEGEL M. (eds), *Les Formes du Sens*, Duculot, Louvain La Neuve, 1996.

[KLÜ 00] KLÜTER A., ALASSANE N., KIRCHMANN H., "Verbmobil from a software engineering point of view: system design and software integration", in WAHLSTER W. (ed.), *Verbmobil: Foundations of Speech-to-Speech Translation*, Springer, Berlin, 2000.

[KLU 05] KLUGE W., *Abstract Computing Machines: A Lambda Calculus Perspective*, Springer-Verlag, Berlin/Heidelberg, 2005.

[KNI 99] KNIGHT K., *A Statistical MT Tutorial Workbook*, prepared in connection with the JHU summer workshop, available at: http://www.isi.edu/natural-language/mt/wkbk.rtf, April 30, 1999.

[KOE 05] KOEHN P., "Europarl: a parallel corpus for statistical machine translation", *The Tenth Machine Translation Summit*, Phuket, Thailand, 12–16 September 2005.

[KOH 97] KOHONEN T., *Self-Organizing Maps*, 2nd extended edition, Springer, 1997.

[KOH 99] KOHAVI R., QUINLAN R.J., "Decision tree discovery", in KLOSGEN W., ZYTKOW J.M. (eds), *Handbook of Data Mining and Knowledge Discovery*, Oxford University Press, Oxford, 1999.

[KOW 11] KOWALSKY G., *Information Retrieval Architecture and Algorithms*, Springer, New York, 2011.

[KRA 00] KRAHMER E., PIWEK P., *Varieties of Anaphora*, 12th ESSLLI Summer School, Birmingham, 2000, available at: http://citeseerx.ist.psu.edu/viewdoc/summary?doi=10.1.1.34.4525, 2000.

[KRE 97] KRENN B., SAMUELSSON C., *Statistical Methods in Computational Linguistics*, Esslli Summerschool, Aix-en-Provence, 1997.

[KUR 01] KURDI M.-Z., "A spoken language understanding approach which combines the parsing robustness with the interpretation deepness", *The International Conference on Artificial Intelligence IC-AI01*, Las Vegas, NV, 25–28 June 2001.

[KUR 02] KURDI M.-Z., AHAFHAF M., "Toward an objective and generic method for spoken language understanding systems evaluation: an extension of the DCR method", *Third International Conference on Language Resources and Evaluation LREC02*, Las Palmas, 2002.

[KUR 11] KURDI M.-Z., "Personalized language learning through adaptive Computer software: application to French", *International Conference for Computer Applications ICCA 2011*, Riyadh, Saudi Arabia, 31 May–2 June 2011.

[KUR 16] KURDI M.-Z., *Natural Language Processing and Computational Linguistics: Speech, Morphology and Syntax*, ISTE Ltd, London and John Wiley & Sons, New York, 2016.

[KWA 08] KWAK S., AOYAMA T., Coreference resolution with decision tree, Technical report CS224N, Final Project Stanford University, Spring 2008.

[LAB 04] LABELLE F., "La logique des langues naturelles, Cours de sémantique", University of Quebec, available at: wwwens.uqac.ca/~flabelle/semantique/, 2004.

[LAC 97] LACHENAUD G., "Discours et récit chez les historiens grecs : l'apport des théories de Benveniste", *Linx* [E-book], published 3 July 2012, accessed 9 May 2015, available at: http://linx.revues.org/1040, 9 | 1997. doi: 10.4000/linx.1040.

[LAF 94] LAFOURCADE M., Génie Logiciel pour le Génie Linguiciel, PhD Thesis, Joseph Fourier University, 1994.

[LAK 80] LAKOFF G., JOHNSON M., *Metaphors We Live By*, University of Chicago, Chicago, IL, 1980.

[LAK 87] LAKOFF G., *Women, Fire, and Dangerous Things: What Categories Reveal About the Mind*, University of Chicago Press, Chicago, 1987.

[LAK 89] LAKOFF G., TURNER M., *More than Cool Reason: A Field Guide to Poetic Metaphor*, Chicago University Press, Chicago, 1989.

[LAN 87] LANGACKER R.W., *Foundations of Cognitive Grammar*, vol. 1, *Theoretical Prerequisites*, Stanford University Press, Stanford, 1987.

[LAN 97] LANDAUER T.K., DUMAIS S.T., "A solution to Plato's problem: the latent semantic analysis theory of the acquisition, induction, and representation of knowledge", *Psychological Review*, vol. 104, pp. 211–240, 1997.

[LAN 98] LANDAUER T.K., FOLTZ P.W., LAHAM D., "Introduction to latent semantic analysis", *Discourse Processes*, vol. 25, pp. 259–284, available at: http://lsa.colorado.edu/, 1998.

[LAN 05] LAN M., TAN C.L., LOW H.B. *et al.*, "A comprehensive comparative study on term weighting schemes for text categorization with support vector machines", *14th International World Wide Web Conference*, Chiba, Japan, 10–14 May 2005.

[LAU 99] LAURENCE S., MARGOLIS E., "Concepts and cognitive science", in MARGOLIS E., LAURENCE S. (eds), *Concepts: Core Readings*, MIT Press, Cambridge, 1999.

[LEC 06] LECHNER W., "Introduction to formal semantics", available at: http://vivaldi.sfs.nphil.uni-tuebingen.de/%7Ennsle01/Introsem.htm, 2006.

[LEI 03] LEIDNER J.L., "Current issues in software engineering for natural language processing", *Proceedings of the HLT-NAACL Workshop on Software Engineering and Architecture of Language Technology Systems*, vol. 8, 2003.

[LET 04] LETE B., SPRENGER-CHAROLLES L., COLE P., "MANULEX: a grade-level lexical database from French elementary-school readers", *Behavior Research Methods, Instruments & Computers*, vol. 36, pp. 156–166, 2004.

[LEV 04] LEVOW G.-A., "Prosodic cues to discourse segment boundaries in human-computer dialogue", *The 5th SIGdial Workshop on Discourse and Dialogue*, Boston, MA, 30 April and 1 May, 2004.

[LEW 70] LEWIS D., "General semantics", *Synthese*, vol. 22, pp. 18–67, 1970.

[LEW 72] LEWIS D., "General semantics", in DAVIDSON D., HARMAN G. (eds), *Semantics of Natural Language*, D. Reidel Publishing Company, Dordrecht, 1972.

[LEW 05] LEWIS W.E., *Software Testing and Continuous Quality Improvement*, Auerbach Publications, New York, 2005.

[LIU 11] LIU B., *Web Data Mining Exploring Hyperlinks, Hyperlinks, Contents and Usage Data*, 2nd ed., Springer, 2011.

[LLO 82] LLOYD S.P., "Least squares quantization in PCM", *IEEE Transactions on Information Theory*, vol. 28, no. 2, pp. 129–137, 1982.

[LOC 98] LOCHBAUM K., "A collaborative planning model of intentional structure", *Computational Linguistics*, vol. 24, no. 4, pp. 525–572, 1998.

[LOD 02] LODHI H., SAUNDERS C., SHAWE-TAYLOR J. *et al.*, "Text classification using string kernels", *Journal of Machine Learning Research*, vol. 2, pp. 419–444, available at: http://www.jmlr.org/papers/volume2/lodhi02a/lodhi02a.pdf, 2002.

[LOP 08] LOPEZ A., "Statistical machine translation", *ACM Computing Surveys (CSUR)*, vol. 40, no. 3, p. 8, 2008.

[LÖV 12] LÖVEHEIM H., "A new three-dimensional model for emotions and monoamine neurotransmitters", *Med Hypotheses*, vol. 78, no. 2, pp. 341–348, 2012.

[LUT 06] LUTZ M., *Programming Python*, 3rd ed., O'Reilly, 2006.

[LUZ 95] LUZZATTI D., *Le dialogue verbal homme-machine : étude de cas*, Masson, Paris, 1995.

[LYC 00] LYCAN W., *Philosophy of Language: A Contemporary Introduction*, Routledge, London, 2000.

[LYO 66] LYONS J., "Towards a 'notional' theory of the 'parts of speech'", *Journal of Linguistics*, vol. 2, pp. 209–236, 1966.

[MAC 81] MACKENZIE J.L., "Pedagogically relevant aspects of case grammar", in JAMES A., WESTNEY P. (eds), *New Linguistic Impulses in Foreign Language Teaching*, Gunter Narr Verlag, Tubingen, 1981.

[MAE 03] MAEDCH A.D., *Ontology Learning for the Semantic Web*, Kluwer Academic Publishers, Norwell, 2003.

[MAH 95] MAHESH K., Syntax semantic interaction in sentence understanding, PhD Dissertation, Georgia Institute of Technology, 1995.

[MAL 06] MALIOUTOV I., BARZILAY R., "Minimum cut model for spoken lecture segmentation", *Proceedings of ACL*, pp. 25–32, 2006.

[MAN 57] MANDELBROT B., *Étude de la loi d'Estoup et de Zipf : fréquences des mots dans le discours, dans Logique, langage et théorie de l'information*, PUF, Paris, 1957.

[MAN 84] MANDLER J.M., *Stories, Scripts, and Scenes: Aspects of Schema Theory*, Lawrence Erlbaum Associates, Hillsdale, 1984.

[MAN 88] MANN W.C., THOMPSON S.A., "Rhetorical structure theory: toward a functional theory of text organization", *Text*, vol. 8, no. 3, pp. 243–281, 1988.

[MAN 06] MANGEOT M., "Papillon project: retrospective and perspectives", in ZWEIGENBAUM P. (ed.), *Proceedings of Acquiring and Representing Multilingual, Specialized Lexicons: the Case of Biomedicine, LREC workshop*, Genoa, Italy, 22 May, 2006.

[MAN 08] MANNING C.D., RAGHAVAN P., SCHÜTZE H., *An Introduction to Information Retrieval*, Cambridge University Press, Cambridge, 2008.

[MAN 12] MANN W.C., TABOADA M., "Introduction à la théorie de la structure rhétorique", Simon Fraser University, available at: http://www.sfu.ca/rst/07french/index.html, 2012.

[MAN 14] MANOOCHEHRI M., *Data Just Right: Introduction to Large Scale Data and Analytics*, Addison Wesley, Upper Saddle River, 2014.

[MAR 70] MARTINET A., *Eléments de linguistique générale*, Armand Colin, Paris, 1970.

[MAR 91] MARTIN R., "Typicalité et sens des mots", in DUBOIS D. (ed.), *Sémantique et cognition : catégories, prototypes, typicalité*, Éditions du CNF, Paris, 1991.

[MAR 00] MARCU D., "The rhetorical parsing of unrestricted texts: a surface-based approach", *Computational Linguistics*, vol. 26, pp. 395–448, 2000.

[MAR 01] MARTIN J.R., "Cohesion and texture", in SCHIFFRIN D., TANNEN D., HAMILTON H. (eds), *The Handbook of Discourse Analysis*, Blackwell, Malden, 2001.

[MAS 03] MASOLO C., BORGO S., GANGEMI A. *et al.*, "WonderWeb Deliverable D17: The WonderWeb Library of Foundational Ontologies Preliminary Report", Intermediate Report, Version 2.1, ISTC-CNR, available at: http://www.loa.istc.cnr.it/old/Papers/DOLCE2.1-FOL.pdf, 2003.

[MAS 07] MASCARDI V., ROSSO P., CORDI V., "Enhancing communication inside multi-agent systems: an approach based on alignment via upper ontologies", *Proc. Int'l Workshop Agents, Web-Services and Ontologies: Integrated Methodologies*, 2007.

[MAU 08] MAUREL D., "Prolexbase: a multilingual relational lexical database of proper names", *Proceedings of the Language Resources and Evaluation conference (LREC)*, Marrakech, 28–30 May 2008.

[MCC 76] MCCAWLEY J., *Grammar and Meaning*, Academic Press, New York, 1976.

[MCC 95] MCCARTHY J.F., LEHNERT W.G, "Decision trees for coreference resolution", *The Fourteenth International Conference on Artificial Intelligence IJCAI*, pp. 1050–1055, 1995.

[MCE 96] MCENERY T., WILSON A., *Corpus Linguistics*, Edinburgh University Press, Edinburgh, 1996.

[MEG 03] MEGERDOOMIAN K., "Text mining, corpus building and testing", in FARGHALI A. (ed.), *Handbook for Language Engineers*, Center for the Study of Language and Information, Stanford 2003.

[MEL 84] MEL'CUK I., ARBATCHEWSKY-JUMARIE N., ELNITSKY L. *et al.*, *Dictionnaire explicatif et combinatoire du français contemporain : Recherches lexicosémantiques I*, PUM, Montreal, 1984.

[MEL 97] MEL'CUK I., Vers une linguistique Sens-Texte, Leçon inaugurale, Collège de France, Paris, 1997.

[MEL 98] MEL'CUK I., "The meaning-text approach to the study of natural language and linguistic functional models", in EMBLETON S. (ed.), *LACUS Forum 24*, LACUS, Chapel Hill, pp. 3–20, 1998.

[MEL 99a] MELAMED I.D., "Bitext maps and alignments via pattern recognition", *Computational Linguistics*, vol. 25, no. 1, pp. 107–130, 1999.

[MEL 99b] MELBY A.K., "SALT: standards-based access service to multilingual lexicons and terminologies", available at: http:// www.ttt.org, 1999.

[MER 03] MERTZ D., *Text Processing in Python*, Addison Wesley, Boston, 2003.

[MET 05] METZNER C., CORTEZ L., CHACÍN D., "Using a blackboard architecture in a web application", *Issues in Informing Science and Information Technology IISIT*, vol. 2, 2005.

[MEY 04] MEYER C., *English Corpus Linguistics: An Introduction*, Cambridge University Press, Cambridge, 2004.

[MIC 89] MICHAELSON G., *Introduction to Functional Programming Through Lambda Calculus*, Addison Wesley, 1989.

[MIC 91] MICHIELS A., *Traitement du langage naturel et Prolog : une introduction*, Hermes, Paris, 1991.

[MIC 01] MICHARD A., *XML : langage et application*, Eyrolles, Paris, 2001.

[MID 03] MIDGLEY T.D., "Discourse chunking a tool in dialogue act tagging", *The 41st Annual Meeting on Association for Computational Linguistics ACL '03*, Sapporo, Japan, 2003.

[MIK 13] MIKOLOV T., CHEN K., CORRADO G. *et al.*, "Efficient estimation of word representations in vector space", arXiv:1301.3781, available at: https://arxiv.org/abs/1301.3781, 2013.

[MIL 90] MILLER G.A., BECKWITH R., FELLBAUM C. *et al.*, "Introduction to WordNet: an online lexical database", *International Journal of Lexicography*, vol. 3, no. 4, pp. 235–244, 1990.

[MIL 00] MILTSAKAKI E., KUKICH K., "The role of centering theory's rough-shift in the teaching and evaluation of writing skills", *The 38th Annual Meeting on Association for Computational Linguistics*, pp. 408–4150, 2000.

[MIL 03] MILLER T., "Essay assessment with latent semantic analysis", *Journal of Educational Computing Research*, vol. 29, no. 4, pp. 495–512, 2003.

[MIL 04] MILTSAKAKI E., PRASAD R., JOSHI A. *et al.*, "The Penn discourse treebank", *Proceedings of the NAACL/HLT Workshop on Frontiers in Corpus Annotation*, 2004.

[MIN 75] MINSKY M., "A framework for representing knowledge", in WINSTON P. (ed.), *The Psychology of Computer Vision*, McGraw-Hill, New York, 1975.

[MIN 88] MINSKY N.H., Law-governed systems, Report, Recherche de l'Université de Rutgers, February, 1988.

[MIN 95] MINKER W., An English version of LIMSI L'ATIS system, Technical report, Laboratoire LIMSI no. 95-12, 1995.

[MIN 96] MINKER W., BENNACEF S., "Compréhension et évaluation dans le domaine ATIS", *Journée Francophone JEP96*, 1996.

[MIN 07] MINSKY M., The *Emotion Machine: Commonsense Thinking, Artificial Intelligence, and the Future of the Human Mind*, Simon & Schuster, 2007.

[MIT 76] MITTERAND H., *Les mots français*, PUF, Paris, 1976.

[MIT 94] MITCHELL M., KIM G., MARCINKIEWICZ M.A. *et al.*, "The Penn treebank: annotating predicate argument structure", *Proceedings of the Human Language Technology Workshop*, San Francisco, CA, March, 1994.

[MIT 97] MITKOV R., "Two engines are better than one: generating more power and confidence in the search for the antecedent", in MITKOV R., NICOLOV N. (eds), *Recent Advances in Natural Language Processing*, John Benjamin Publishers, Philadelphia, 1997.

[MIT 98] MITKOV R., "Anaphora resolution: the state of the art", *COLING'98/ACL'98 Tutorial on Anaphora Resolution*, 1998.

[MON 70] MONTAGUE R., "English as a Formal Language", in VISENTINI B. (ed.), *Linguaggi nella società e nella tecnica*, Mailand, 1970.

[MON 73] MONTAGUE R., "The proper treatment of quantification in ordinary English", in HINTIKKA J., MORAVCSIK J., SUPPES P. (eds), *Approaches to Natural Language*, Reidel, Dordrecht, 1973.

[MON 00] MONTES-Y-GÓMEZ M., LÓPEZ-LÓPEZ M.A., GELBUKH A., "Information retrieval with conceptual graph matching", *DEXA'00 Proceedings of the 11th International Conference on Database and Expert Systems Applications*, London, 2000.

[MOO 93] MOORE J., PARIS C., "Planning text for advisory dialogues: capturing intentional and rhetorical information", *Computational Linguistics*, vol. 19, no. 4, pp. 651–695, 1993.

[MOO 02] MOORE R.C., "Fast and accurate sentence alignment of bilingual corpora", *Machine Translation: From Research to Real Users, 5th Conference of the Association for Machine Translation in the Americas*, Springer-Verlag, Heidelberg, Germany, pp. 135–244, 2002.

[MÜL 00] MÜLLER S., KASPER W., "HPSG analysis of German", in WAHLSTER W. (ed.), *Verbmobil: Foundations of Speech-to-Speech Translation*, Springer, Berlin, 2000.

[MYE 00] MYERS K., KEARNS M.J., SINGH S.P. *et al.*, "A boosting approach to topic spotting on subdialogues", *Proceedings of the Seventeenth International Conference on Machine Learning ICML '00*, 2000.

[NAD 06] NADEAU D., TURNEY P., MATWIN S., "Unsupervised named entity recognition: generating gazetteers and resolving ambiguity", *The 19th Canadian Conference on Artificial Intelligence*, Quebec City, Quebec, Canada, 7–9 June 2006.

[NAD 07] NADEAU D., SEKINE S., "A survey of named entity recognition and classification", *Lingvisticæ Investigationes*, vol. 30, no. 1, pp. 3–26, available at: http://brown.cl.uni-heidelberg.de/~sourjiko/NER_Literatur/survey.pdf, 2007.

[NAG 84] NAGAO M., "A framework of a mechanical translation between Japanese and English by analogy principle", in ELITHORN A., BANERJI R. (eds), *Artificial and Human Intelligence*, Elsevier Science Publishers, available at: http://www.mt-archive.info/Nagao-1984.pdf, 1984.

[NAN 04] NANAS N., UREN V., DE ROECK A., "A comparative study of term weighting methods for information filtering", in GALINDO F., TAKIZAWA M., TRAUNMÜLLER R. (eds), *Proceedings of the 15th International Workshop on Database and Expert Systems Applications*, IEEE Computer Society, pp. 13–17, available at: http://kmi.open.ac.uk/publications/papers/kmi-tr-128.pdf, 2004.

[NAS 90] NASRI M.-K., Architecture du système de reconnaissance automatique de la parole DIRA, PhD Thesis, National Polytechnic Institute of Grenoble, 1990.

[NES 05] NESSELHAUF N., *Collocations in a Learner Corpus*, John Benjamins Publishing Company, 2005.

[NEW 01] NEW B., PALLIER C., FERRAND L. *et al.*, "Une base de données lexicales du français contemporain sur Internet : LEXIQUE", *L'Année Psychologique*, vol. 101, pp. 447–462, 2001.

[NEW 04] NEW B., PALLIER C., BRYSBAERT M. *et al.*, "Lexique 2: a new French lexical database", *Behavior Research Methods, Instruments, & Computers*, vol. 36, no. 3, pp. 516–524, 2004.

[NEW 06] NEW B., "Lexique 3 : une nouvelle base de données lexicales", *Actes de la Conférence Traitement Automatique des Langues Naturelles (TALN 2006)*, Louvain, Belgium, April, 2006.

[NIL 03] NILES I., PEASE A., "Linking lexicons and ontologies: mapping WordNet to the suggested upper merged ontology", *Proceedings of the IEEE International Conference on Information and Knowledge Engineering*, pp. 412–416, 2003.

[OCH 12] OCHS M., SADEK D., PELACHAUD C., "A formal model of emotions for an empathic rational dialog agent", *Autonomous Agents and Multi-Agent Systems Archive*, vol. 24, no. 3, pp. 410–440, 2012.

[ODO 08] O'DONNELL M., "Demonstration of the UAM CorpusTool for text and image annotation", *Proceedings of the ACL-08:HLT Demo Session (CompanionVolume)*, Columbus, OH, Association for Computational Linguistics, pp. 13–16, June 2008.

[OGD 23] OGDEN C.K., RICHARD I.A., *The Meaning of Meaning*, Routledge and Kegan Paul, London, 1923.

[OLM 08] OLMOS ALBACETE R., LEON J.A., JORGE-BOTANA G., "Using latent semantic analysis vs. human judgements assessing short summaries in expository texts", available at: http://www.researchgate.net/publication/267836513_Using_Latent_Semantic_Analysis_vs._Human_Judgements_assessing_short_summaries_in_expository_texts, 2008.

[ONO 94] ONO K., SUMITA K., MIIKE S., "Abstract generation based on rhetorical structure extraction", *Proceedings, 15th International Conference on Computational Linguistics COLING*, Kyoto, Japan, pp. 344–348, 5–9 August 1994.

[ORT 87] ORTONY A., CLORE G.L., FOSS M.A., "The psychological foundations of the affective lexicon", *Journal of Personality and Social Psychology*, vol. 53, pp. 751–766, 1987.

[ORT 88] ORTONY A., CLORE G.L., COLLINS A., *The Cognitive Structure of Emotions*, Cambridge University Press, Cambridge, 1988.

[OSG 57] OSGOOD C.E., SUCI G., TANNENBAUM P., *The Measurement of Meaning*, University of Illinois Press, Urbana, 1957.

[OSG 75] OSGOOD C.E., MAY W.H., MIRON M.S., *Cross-Cultural Universals of Affective Meaning*, University of Illinois Press, Urbana, 1975.

[OUN 98] OUNIS I., Un modèle d'indexation relationnel pour les graphes conceptuels fondé sur une interprétation logique, PhD Thesis, Joseph Fourier University, Grenoble, 1998.

[PAC 91] PACHERIE E., "Aristote et Rosch un air de famille ?", in DUBOIS D. (ed.), *Sémantique et cognition : catégories, prototypes, typicalité*, Éditions du CNF, Paris, 1991.

[PAK 12] PAK A., Automatic, adaptive, and applicative sentiment analysis, PhD Thesis, University of Paris-Sud, available at: https://tel.archives-ouvertes.fr/tel-00717329/document, 2012.

[PAL 05] PALMER M., DANIEL G., KINGSBURY P., "The proposition bank: an annotated corpus of semantic roles", *Computational Linguistics*, vol. 31, no. 1, pp. 71–106, 2005.

[PAL 10] PALMER M., XUE N., "Linguistic annotation", in CLARK A., FOX C., LAPPIN S. (eds), *The Handbook of Computational Linguistics and Natural Language Processing*, Wiley-Balckwell, Malden, 2010.

[PAN 08] PANG B., LEE L., "Opinion mining and sentiment analysis", *Foundations and Trends in Information Retrieval*, vol. 2, nos. 1–2, pp. 1–135, 2008.

[PAR 02] PARTEE B., "Noun phrase interpretation and type-shifting principles", in PORTNER P., PARTEE B. (eds), *Formal Semantics: the Essential Readings*, Blackwell, Oxford, 2002.

[PAR 07] PAROUBEK P., CHAUDIRON S., HIRSCHMAN L., "Principles of evaluation in natural language processing", *TAL*, vol. 48, no. 1, pp. 7–31, 2007.

[PAS 08] PASCA M., VAN DURME B., "Weakly-supervised acquisition of open-domain classes and class attributes from web documents and query logs", *Proceedings of ACL-08: HLT*, pp. 19–27, 2008.

[PAT 93] PATRY R., "L'analyse du niveau discursive en linguistique", in NESPOULOUS J.-L. (ed.), *Tendances actuelles en linguistique générale*, Delachaux et Niestlée, Paris, 1993.

[PAU 99] PAUL M., KYAMAMOTO A., SUMITA E., "Corpus-based anaphora resolution towards antecedent preference", *The Workshop on Coreference and its Applications*, pp. 47–52, 1999.

[PEA 02] PEASE A., NILES I., LI J., The suggested upper merged ontology: a large ontology for the semantic web and its applications, AAAI Technical Report WS-02-11, 2002.

[PEA 09] PEASE A., "Standard upper ontology knowledge interchange format", available at: http://sigmakee.cvs.sourceforge.net/*checkout*/sigmakee/sigma/suo-kif.pdf, 2009.

[PEE 99] PEEREMAN R., CONTENT A., "Lexop: a lexical database with orthography-phonology statistics for French monosyllabic words", *Behavior Research Methods, Instruments, & Computers*, vol. 31, no. 2, pp. 376–379, 1999.

[PEL 00] PELLOM B., WARD W., PRADHAN S., "The CU communicator: an architecture for dialogue systems", *International Conference on Spoken Language Processing (ICSLP'00)*, Beijing, China, November, 2000.

[PEL 01] PELLOM B., WARD W., HANSEN J. *et al.*, "University of Colorado Dialog Systems for travel and navigation", *Human Language Technology Conference (HLT-2001)*, San Diego, CA, March 2001.

[PER 87] PEREIRA F., SHIEBER S., *Prolog and Natural Language Analysis*, CSLI, Stanford, CA, 1987.

[PER 94] PERAKATH C.B., MENZEL C.P., MAYER R.J. *et al.*, *The IDEF5 Ontology Description Capture Method Overview*, Knowledge Based Systems, 1994.

[PER 97] PERRAMOND D.B., *Composition et grammaire de texte*, Canadian Scholar's Press, Toronto, 1997.

[PIC 77] PICOCHE J., *Précis de lexicologie française*, Nathan, Paris, 1977.

[PIC 86] PICOCHE J., *Structures sémantiques du lexique français*, Nathan, Paris, 1986.

[PLU 01] PLUTCHIK R., "The nature of emotions", *American Scientist*, vol. 89, no. 4, pp. 344–350, 2001.

[POE 98] POESIO M., RENATA V., "A corpus-based investigation of definite description use", *Computational Linguistics*, vol. 24, no. 2, pp. 183–216, 1998.

[POE 99] POESIO M., HENSCHEL R., HITZEMAN J. *et al.*, "Towards an annotation scheme for noun phrase generation", *Proceedings of the EACL Workshop on Linguistically Interpreted Corpora*, Bergen, June, 1999.

[POE 04a] POESIO M., "Discourse annotation and semantic annotation in the GNOME corpus", *Proceedings of the ACL Workshop on Discourse Annotation*, pp. 72–79, 2004.

[POE 04b] POESIO M., "The MATE/GNOME Scheme for Anaphoric Annotation", *Proceedings of SIGDIAL*, Boston, MA, April 2004.

[POL 98] POLGUERE A., "La théorie Sens-Texte", *Dialangue*, vols. 8–9, Université du Québec à Chicoutimi, pp. 9–30, 1998.

[POL 00] POLGUERE A., "Towards a theoretically-motivated general public dictionary of semantic derivations and collocations for French", *Proceeding of EURALEX'2000*, Stuttgart, pp. 517—527, 2000.

[POL 01] POLANYI L., "The linguistic structure of discourse", in SCHIFFRIN D., TANNEN D., HAMILTON H.E. (eds), *The Handbook of Discourse Analysis*, Blackwell, Oxford, 2001.

[POL 04a] POLANYI L., CULY C., VAN DEN BERG M. *et al.*, "A rule-based approach to discourse parsing", *Proceedings of the 5th SIGdial Workshop on Discourse and Dialogue*, ACL, Stroudsburg, PA, 2004.

[POL 04b] POLANYI L., CULY C., VAN DEN BERG M. *et al.*, "Sentential structure and discourse parsing", *The ACL Workshop on Discourse Annotation*, Barcelona, Spain, 2004.

[POL 04c] POLANYI L., ZAENEN A., "Contextual valence shifters", *AAAI Spring Symposium on Attitude*, Stanford, CA, p. 10, 2004.

[POT 64] POTTIER B., "Vers une sémantique moderne", *Travaux de linguistique et de littérature*, vol. 2, pp. 107–138, 1964.

[POW 03] POWERS S., *Practical RDF*, O'Reilly, Cambridge, 2003.

[PRI 81] PRINCE E.F., "Toward a taxonomy of given new information", in PRINCE E.F. (ed.), *Syntax and Semantics: Vol. 14. Radical Pragmatics*, Academic Press, New York, 1981.

[PRI 92] PRINCE E.F., "The ZPG letter subjects deniteness and information status", in THOMPSON S., MANN W. (eds), *Discourse Description Diverse Analyses of a Fundraising Text*, John Benjamins, 1992.

[PRI 94a] PRIGENT G., "Synchronous TAGs and machine translation", *Proceedings of the Third International TAG Workshop*, Paris, 1994.

[PRI 94b] PRINCE E.F., "The notion of construction and the syntax discourse interface", *The 25th Annual Meeting of the North East Linguistic Society*, University of Pennsylvania, 1994.

[PUS 91] PUSTEJOVSKY J., "The generative lexicon", *Computational Linguistics*, vol. 17, no. 4, pp. 409–441, 1991.

[PUS 93] PUSTEJOVSKY J., BOGURAEV B., "Lexical knowledge representation and natural language processing", *Artificial Intelligence*, vol. 63, pp. 193–223, 1993.

[PUS 95] PUSTEJOVSKY J., *The Generative Lexicon*, MIT Press, Cambridge, 1995.

[PUS 03] PUSTEJOVSKY J., HANKS P., SAURI R. *et al.*, "The TimeBank corpus", *Corpus Linguistics*, pp. 647–656, 2003.

[PUS 12] PUSTEJOVSKY J., STUBBS A., *Natural Language Annotation for Machine Learning*, O'Reily, Sebastopol, CA, 2012.

[QUI 68] QUILLIAN M.R., "Semantic memory", in MINSKY M. (ed.), *Semantic Information Processing*, MIT Press, Cambridge, 1968.

[QUI 79] QUINLAN J.R., "Discovering rules from large collections of examples: a case study", in MICHIE D. (ed.), *Expert Systems in the Micro Electronic Age*, Edinburgh University Press, Edinburgh, 1979.

[QUI 93] QUINLAN J.R., *C4.5: Programs for Machine Learning*, Morgan Kaufmann, San Mateo, 1993.

[RAB 10] RABATEL A., "Retour sur les relations entre locuteur et énonciateur : Des voix et des points de vue", in COLAS-BLAISE M., KARA M., PERRIN L. *et al.* (eds), *La question polyphonique ou dialogique dans les sciences du langage*, CELTED, University of Metz, 2010.

[RAJ 16] RAJENDRAN P., BOLLEGALA D., PARSONS S., "Contextual stance classification of opinions: A step towards enthymeme reconstruction in online reviews", *Proceedings of the 3rd Workshop on Argument Mining*, Berlin, Germany. Association for Computational Linguistics, pp. 31–39, available at: https://aclweb.org/anthology/W/W16/W16-2804.pdf, 7–12 August 2016.

[RAS 73] RASTIER F., *Essais de sémiotique discursive*, Mame, Paris, E-book available at: http://www.revue-texto.net/Parutions/Essais-de-semiotique/Rastier_essais_de_semiotique.html, 1973.

[RAS 89] RASTIER F., *Sens et textualité*, Hachette, Paris, 1989.

[RAS 90] RASTIER F., "La triade sémiotique, le trivium et la sémiotique linguistique", *Nouveaux actes sémiotiques*, no. 9, pp. 5–39, 1990.

[RAS 91a] RASTIER F., "Catégorisation, typicalité et lexicologie", in DUBOIS D. (ed.), *Sémantique et cognition : catégories, prototypes, typicalité*, Éditions du CNF, Paris, 1991.

[RAS 91b] RASTIER F., *Sémantique et recherches cognitives*, PUF, Paris, 1991.

[RAS 94] RASTIER F., *Sémantique pour l'analyse*, Masson, Paris, 1994.

[RAS 01] RASTIER F., *Arts et sciences du texte*, PUF, Paris, 2001.

[RAS 02] RASTIER F., "La macrosémantique", *Texto !*, available at: http://www.revue-texto.net/Inedits/Rastier/Rastier_Marcosemantique1.html, June 2002.

[RAS 05a] RASTIER F., "La microsémantique", *Texto !*, vol. X, no. 2, available at: http://www.revue-texto.net/Inedits/Rastier/Rastier_Microsemantique.html, June 2005.

[RAS 05b] RASTIER F., "Mésosémantique et syntaxe", *Texto !* [e-book], available at: http://www.revue-texto.net/Inedits/Rastier/Rastier_Mesosemantique.html, September 2005.

[RAY 83] RAYNER K., CARLSON M., FRANSIER L., "The interaction of syntax-semantics during sentence processing: eye movement in the analysis of sentence analysis", *Journal of Verbal learning and Verbal Behavior*, vol. 22, pp. 358–374, 1983.

[RAY 92] RAYNER K., GARROD S., PERFETTI C.A., "Discourse influences during parsing are delayed", *Cognition*, vol. 45, pp. 109–139, 1992.

[RAY 03] RAY E., *Learning XML*, 2nd ed., O'Reilly, 2003.

[RES 10] RESNIK P., LIN J., "Evaluation of NLP systems, in Alexander CLARK", in FOX C., LAPPIN S. (eds), *The Handbook of Computational Linguistics and Natural Language Processing*, Wiley-Balckwell, Malden, 2010.

[RIG 99] RIGAU I., CLARAMUNT G., *Ontologies, Automatic Acquisition of Lexical Knowledge from MRDS*, Politécnica de Catalunya, 1999.

[RIL 06] RILOFF E., PATWARDHAN S., WIEBE J., "Feature subsumption for opinion analysis", *Proceedings of the Conference on Empirical Methods in Natural Language Processing (EMNLP-06)*, 2006.

[RIT 12] RITTER A., SAM M., ETZIONI O. *et al.*, "Open domain event extraction from Twitter", *The 18th ACM SIGKDD International Conference on Knowledge Discovery and Data Mining*, pp. 1104–1112, 2012.

[RIV 11] RIVIÈRE J., ADAM C., PESTY S. *et al.*, "Expressive multimodal conversational acts for SAIBA agents", *Intelligent Virtual Agents*, Reykjavik, Iceland, Lecture Notes in Computer Science, Springer, 6895, pp. 316–323, 2011.

[ROB 67] ROBERT P., REY A., REY-DEBOVE J. *et al.*, *Le Petit Robert : dictionnaire de la langue française*, 2nd ed., Société du nouveau Littré/Le Robert, Paris, 1967.

[ROB 76] ROBERTSON S.E., SPÄRCK JONES K., "Relevance weighting of search terms", *Journal of the American Society for Information Science*, vol. 27, no. 3, pp. 129–146, 1976.

[ROB 04] ROBERTSON S.E., "Understanding inverse document frequency: on theoretical arguments for IDF", *Journal of Documentation*, vol. 60, no. 5, pp. 503–520, 2004.

[ROJ 98] ROJAS R., *A Tutorial Introduction to the Lambda Calculus*, FU Berlin, WS-97/98, available at: www.utdallas.edu/~gupta/courses/apl/lambda.pdf, 1998.

[ROS 73] ROSCH E., "Natural categories", *Cognitive Psychology*, vol. 4, pp. 328–350, 1973.

[ROS 75a] ROSCH E., MERVIS C.B., "Family resemblances: studies in the internal structure of categories", *Cognitive Psychology*, vol. 7, pp. 573–605, 1975.

[ROS 75b] ROSCH E., "Cognitive representations of semantic categories", *Journal of Experimental Psychology: General*, vol. 104, pp. 192–233, 1975.

[ROS 76] ROSCH E., MERVIS C.B., GRAY W. *et al.*, "Basic objects in natural categories", *Cognitive Psychology*, vol. 8, pp. 382–439, 1976.

[ROS 78] ROSCH E., "Principles of categorization", in ROSCH E., LLOYD B.B. (eds), *Cognition and Categorization*, Erlbaum, Hillsdale, 1978.

[ROS 06] ROSELL M., *Introduction to Information Retrieval And Text Clustering*, KTH CSC, August, 2006.

[ROU 00] ROUILLARD J., Hyperdialogue sur Internet : Le système HALPIN, PhD Thesis, University of Grenoble I, 2000.

[RUL 00] RULAND T., "Probabilistic LR-parsing with symbolic postprocessing", in WAHLSTER W. (ed.), *Verbmobil: Foundations of Speech-to-Speech Translation*, Springer, Berlin, 2000.

[RUM 05] RUMBAUGH J., JACOBSON I., BOOCH G., *The Unified Modeling Language Reference Manual*, 2nd ed., Addison-Wesley, 2005.

[RUP 00] RUPP C.J., SPIKLER J., KLARNER M. *et al.*, "Combining analyses from various parses", in WAHLSTER W. (ed.), *Verbmobil: Foundations of Speech-to-Speech Translation*, Springer, Berlin, 2000.

[RUS 80] RUSSEL J., "A circumplex model of affect", *Journal of Personality and Social Psychology*, vol. 39, no. 6, pp. 1161–1178, 1980.

[RUS 95] RUSSEL S., NORVIG P., *Artificial Intelligence: A Modern Approach*, Prentice Hall, New Jersey, 1995.

[RUS 14a] RUSSELL M.A., *Mining the Social Web*, 2nd ed., O'Reilly, Sebastopol, 2014.

[RUS 14b] RUSSELL J., *Agile Data Science*, O'Reilly, Sebastopol, 2014.

[SAB 90] SABAH G., "CARAMEL : un système multi-experts pour le traitement automatique des langues", *Modèles linguistiques*, vol. 12, no. 1, pp. 95–118, 1990.

[SAG 08] SAGOT B., FIŠER D., "Construction d'un wordnet libre du français à partir de ressources multilingues", *TALN 2008*, Avignon, France, 2008.

[SAH 96] SAHAMI M., "Learning limited dependence Bayesian classifiers", *Proceedings of KDD-96*, pp. 335–338, 1996.

[SAH 98a] SAHAMI M., Using machine learning to improve information access, PhD Thesis, Computer Science Department, Stanford University, 1998.

[SAH 98b] SAHAMI M., DUMAIS S., HECKERMAN D. *et al.*, "A Bayesian approach to filtering junk e-mail", *AAAI'98 Workshop on Learning for Text Categorization*, Madison, WI, 27 July 1998.

[SAL 83] SALTON G., MCGILL M.J., *Introduction to Modern Information Retrieval*, McGraw-Hill, 1983.

[SAL 88] SALTON G., BUCKLEY C., "Term-weighting approaches in automatic text retrieval", *Information Studies and Management*, vol. 24, no. 5, pp. 513–523, 1988.

[SAL 08] SALEM A.-B.M., ALFONSE M., "Ontology versus semantic networks for medical knowledge representation", *ICCOMP'08 Proceedings of the 12th WSEAS International Conference on Computers*, 2008.

[SAN 94] SANDOVAL V., *SGML : un outil pour la gestion électronique de documents*, Hermès, Paris, 1994.

[SCH 75] SCHANK R., ABELSON R., "Scripts, plans, and knowledge", *IJCAI'75 Proceedings of the 4th International Joint Conference on Artificial Intelligence*, Tbilisi, Georgia, USSR, 3–8 September 1975.

[SCH 77] SCHANK R., ABELSON R., *Scripts, Plans, Goals and Understanding: an Inquiry into Human Knowledge Structures*, Lawrence Erlbaum, Hillsdale, 1977.

[SCH 84] SCHERER K.R., "Emotion as a multicomponent process: a model and some cross-cultural data", in SHAVER P. (ed.), *Review of Personality and Social Psychology*, vol. 5, Sage, Beverley Hills, 1984.

[SEA 69] SEARLE J., *Speech Acts*, Cambridge University Press, Cambridge, 1969.

[SEK 07] SEKINE S., ODA A., "System demonstration of on-demand information extraction", *Proceedings of the 45th Annual Meeting of the Association for Computational Linguistics Companion Volume: Proceedings of the Demo and Poster Sessions*, pp. 17–20, 2007.

[SÉR 94] SÉRASSET G., SUBLIM : un système universel de bases lexicales multilingues et NADIA : sa spécialisation aux bases lexicales interlingues par acceptions, PhD Thesis, Joseph Fourier University, Grenoble, France, 1994.

[SÉR 99] SÉRASSET G., BOITET C., *UNL-French Deconversion as Transfer & Generation from an Interlingua with Possible Quality Enhancement through Offline Human Interaction*, MT-Summit, Singapore, pp. 220–228, 1999.

[SÉR 01] SÉRASSET G., MANGEOT M., "Papillon lexical database project: monolingual dictionaries & interlingual links", *NLPRS 2001*, Hitotsubashi Memorial Hall, National Center of Sciences, Tokyo, Japan, 27–30 November 2001.

[SÉR 14] SÉRASSET G., "Dbnary: Wiktionary as a lemon based RDF multilingual lexical resource", *Semantic Web Journal-Special issue on Multilingual Linked Open Data*, 2014.

[SHA 48] SHANNON CLAUDE E., "A mathematical theory of communication", *Bell System Technical Journal*, vol. 27, pp. 379–423, July and October 1948.

[SHI 90] SHIEBER S., SCHABES Y., "Synchronous tree-adjoining grammars", *13th International Conference on Computational Linguistics*, vol. 3, pp. 1–6, 1990.

[SHI 06] SHINYAMA Y., SEKINE S., "Preemptive information extraction using unrestricted relation discovery", *Proceedings of the Human-Language Technology Conference of the NAACL, Main Conference*, pp. 304–311, 2006.

[SID 79] SIDNER C.L., Toward a computational theory of definite anaphora comprehension in English discourse, PhD Dissertation, Massachusetts Institute of Technology, Cambridge, MA, 1979.

[SIE 00] SIEGEL M., "HPSG analysis of English", in WAHLSTER W. (ed.), *Verbmobil: Foundations of Speech-to-Speech Translation*, Springer, Berlin, 2000.

[SIN 91] SINCLAIR J., *Corpus, Concordance, Collocation*, Oxford University Press, Oxford, 1991.

[SLO 81] SLOMAN A., "Why robots will have emotions", *Proceedings International Joint Conference on Artificial Intelligence JCAI'81*, 1981.

[SMI 97] SMITH B., VARZI A., "Fiat and bona fide boundaries", *The Electronic Journal of Analytic Philosophy*, vol. 5, no. 5, available at: http://ejap.louisiana.edu/EJAP/1997.spring/smithvarzi976.html, 1997.

[SMI 04] SMITH B., GRENON P., "The cornucopia of formal-ontological relations", *Dialectica*, vol. 58, no. 3, pp. 279–296, 2004.

[SMI 10] SMITH C., CROOK N., BOYE J. *et al.*, "Interaction strategies for an affective conversational agent", *10th International Conference on Intelligent Virtual Agents*, Philadelphia, PA, USA, September, 2010.

[SMR 03] SMRZ P., POVOLNY M., "DEB – A Dictionary Editor and Browser, Papillon", *Proceedings of the Fourth Papillon Workshop*, Hokkaido University, Sapporo, Japan, 2003.

[SOD 99] SODERLAND S., "Learning information extraction rules for semi-structured and free text", *Machine Learning*, vol. 34, no. 1, pp. 233–272, 1999.

[SOO 01] SOON W.M., NG H.T., LIM D.C.Y., "A machine learning approach to coreference resolution of noun phrases", *Computational Linguistics*, vol. 27, no. 4, pp. 521–544, 2001.

[SOW 76] SOWA J.F., "Conceptual graphs for a database interface", *IBM Journal of Research and Development*, vol. 20, no. 4, pp. 336–357, 1976.

[SOW 83] SOWA J.F., "Generating language from conceptual graphs", *Computer and Mathematics with Applications*, vol. 9, no. 1, pp. 29–43, 1983.

[SOW 86] SOWA J.F., WAY E., "Implementing a semantic interpreter using conceptual graphs", *IBM Journal of Research and Development*, vol. 30, no. 1, pp. 57–69, 1986.

[SOW 92] SOWA J.F., "Semantic networks", in STUART C.S. (ed.), *Encyclopedia of Artificial Intelligence*, 2nd ed., Wiley, 1992.

[SPE 89] SPERBER D., WILSON D., *La Pertinence: Communication Et Cognition*, Éditions de Minuit, Paris, 1989.

[SPI 93] SPIVEY-KNOWLTON M., TANENHAUS M., "Immediate effect of discourse and semantic context in syntactic processing: Evidence from eye tracking", *The Fifteenth Annual Conference of Cognitive Science Society*, pp. 812–817, 1993.

[SPO 05] SPORLEDER C., LAPATA M., "Discourse chunking and its application to sentence compression", *Proceedings of HLT/EMNLP*, 2005.

[STE 95] STEFANINI M.-H., BERRENDONNER A., LALLICH G. *et al.*, "TALISMAN: un système multi-agents gouverné par des lois linguistiques pour le traitement de la langue naturelle", *12th Brazilian Symposium on Artificial Intelligence SBIA*, Campinas, Brazil, pp. 312–322, 1995.

[STE 00] STEINBACH M., KARYPIS G., KUMAR V., "A comparison of document clustering techniques", *Workshop on Text Mining KDD 200*, Boston, MA, USA, 20–23 August 2000.

[STE 04] STEDE M., "The potsdam commentary corpus", *ACL Workshop on Discourse Annotation*, Stroudsburg, PA, 2004.

[STO 03] STONE M., "What is an agent: computational models in artificial intelligence and cognitive science", in LEPORE J., PYLYSHYN Z. (eds), *Cognitive Science*, vol. 2, Blackwell, available at: http://www.cs.rutgers.edu/~mdstone/pubs/whatis.pdf, 2003.

[STR 04] STRAPPARAVA C., VALITUTTI A., "WordNet-affect: an affective extension of WordNet", *Proceedings of the 4th International Conference on Language Resources and Evaluation (LREC)*, Lisbon, pp. 1083–1086, May 2004.

[SUN 92] SUN W. *et al.*, "Agrep: a fast approximate pattern searching tool", *USENIX Conference*, January 1992.

[TET 99] TETRAULT J., "Analysis of syntax-based pronoun resolution methods", *Proceedings of the 37th annual meeting of the Association for Computational Linguistics on Computational Linguistics*, pp. 602–605, 1999.

[THI 04] THIONE G., VAN DEN BERG M., POLANYI L. *et al.*, "Hybrid text summarization: combining external relevance measures with structural analysis", *The ACL 2004 Workshop Text Summarization Branches Out*, Barcelona, Spain, 2004.

[TJO 03] TJONG KIM SANG E., DE MEULDER F., "Introduction to the CoNLL-2003 shared task: language-independent named entity recognition", *Proceedings of the 7th Conference on Natural Language Learning (CoNLL-2003)*, pp. 142–147, 2003.

[TRU 92] TRUSWELL J.C., TANENHAUS M.K., "Consulting temporal context during sentence comprehension: evidence from the monitoring of eye movements in reading", *The Fourteenth Annual Conference of Cognitive Science Society*, pp. 492–497, 1992.

[TRU 04] TRUSTWELL R., Attributive adjectives and the nominal they modify, Master's Dissertation, University of Oxford, 2004.

[TUF 04] TUFIS D., CRISTEA D., STAMOU S., "BalkaNet: Aims, methods, results and perspectives. A general overview.", *Romanian Journal of Information Science and Technology*, vol. 7, nos. 1–2, pp. 9–43, 2004.

[TUR 50] TURING A., "Computing machinery and intelligence", *Mind*, vol. LIX, no. 236, pp. 433–460, 1950. doi: 10.1093/mind/LIX.236.433.

[TUR 01] TURNEY D., "Mining the web for synonyms: PMI-IR versus LSA on TOEFL", *Proceedings of the Twelfth European Conference on Machine Learning*, Freiburg, Germany, pp. 491–502, 2001.

[TYL 77] TYLER L.K., MARSLEN-WILSON W.D., "The on-line effects of semantic context on syntactic processing", *Journal of Verbal Learning and Verbal Behavior*, vol. 16, pp. 683–692, 1977.

[ULR 00] ULRICH H., RULAND T., "Integrated shallow processing", in WAHLSTER W. (ed.), *Verbmobil: Foundations of Speech-to-Speech Translation*, Springer, Berlin, 2000.

[UNL 96] UNL, "Universal Networking Language: an electronic language for communication", *Understanding and Collaboration*, UNL Center, Tokyo, 1996.

[USC 95] USCHOLD M., KING M., "Towards a methodology for building ontologies", *IJCAI-95 Workshop on Basic Ontological Issues in Knowledge Sharing*, Montreal, Canada, 1995.

[USC 96] USCHOLD M., GRÜNINGER M., "Ontologies: principles methods and applications", *Knowledge Engineering Review*, vol. 11, no. 2, pp. 93–136, 1996.

[USZ 00] USZKOREIT H., FLICKINGER D., KASPER W. *et al.*, "Deep linguistic analysis with HPSG", in WOLFGANG W. (ed.), *Verbmobil: Foundations of Speech-to-Speech Translation*, Springer, Berlin, 2000.

[VAN 85] VAN DIJK T.A., "Introduction: discourse analysis as a new cross-discipline", *Journal of Discourse Analysis*, vol. 1, pp. 1–10, 1985.

[VAU 68] VAUQUOIS B., "A survey of formal grammars and algorithms for recognition and transformation in mechanical translation", *IFIP Congress*, no. 2, pp. 1114–1122, 1968.

[VAU 00] VAUFREYDAZ D., BERGAMINI C., SERIGNAT J.-F. *et al.*, "A new methodology for speech corpora definition from internet documents", *Proceedings of the International Conference on Language Resources and Evaluation LREC'2000, 2nd International Conference on Language Resources and Evaluation*, Athens, Greece, vol. 3, pp. 423–426, 2000.

[VET 07] VETULANI Z., WALKOWSKA J., OBREBSKI T. *et al.*, "PolNet – Polish WordNet project algorithm", in VETULANI Z. (ed.), *Proceedings of the 3rd Language and Technology Conference: Human Language Technologies as a Challenge for Computer Science and Linguistics*, Poznan, Poland, Wyd. Poznanskie, Poznan, pp. 172–176, 5–7 October 2007.

[VIL 04] VILLASEÑOR-PINEDA L., MONTES-Y-GÓMEZ M., CAELEN J., "A modal logic framework for human-computer spoken interaction", in GELBUKH A. (ed.), *Computational Linguistics and Intelligent Text Processing*, Lecture Notes in Computer Science, vol. 2945, Springer, 2004.

[VOG 12] VOGT L., GROBE P., QUAST B. *et al.*, "Fiat or bona fide boundary: a matter of granular perspective", *PLoS ONE*, vol. 7, no. 12, p. e48603, 2012. doi: 10.1371/journal.pone.0048603.

[VOL 07] VOLL K., TABOADA M., "Not all words are created equal: extracting semantic orientation as a function of adjective relevance", *20th Australian Joint Conference on Artificial Intelligence*, Gold Coast, Australia, pp. 337–346, December 2007.

[VOO 05] VOORHEES E.M., HARMAN D.K. (eds), *TREC: Experiment and Evaluation in Information Retrieval*, MIT Press, Cambridge, 2005.

[VOS 98] VOSSEN P. (ed.), *EuroWordNet: A Multilingual Database with Lexical Semantic Networks*, Kluwer Academic Publishers, Dordrecht, 1998.

[WAH 95] WAHLSTER W., *Verbmobil: Towards a DRT-based Translation of Spontaneous Negotiation Dialogs*, MT Summit V, Luxembourg, 10–13 July 1995.

[WAH 00a] WAHLSTER W., "Mobile speech-to-speech translation of spontaneous dialogs: an overview of the final Verbmobil system", in WAHLSTER W. (ed.), *Verbmobil: Foundations of Speech-to-Speech Translation*, Springer, Berlin, 2000.

[WAH 00b] WAHLSTER W. (ed.), *Verbmobil: Foundations of Speech-to-Speech Translation*, Springer, Berlin, 2000.

[WAR 05] WARWICK C., "The British National Corpus", available at: http://www.natcorp.ox.ac.uk/, 2005.

[WEA 55] WEAVER W., *Machine Translation of Languages*, MIT Press, Cambridge, 1955.

[WEB 10] WEBBER B., WEBB N., "Question answering", in CLARK A., FOX C., LAPPIN S. (eds), *The Handbook of Computational Linguistics and Natural Language Processing*, Wiley-Balckwell, Malden, 2010.

[WEB 12] WEBBER B., EGG M., KORDONI V., "Discourse structure and language technology", *Natural Language Engineering*, vol. 18, no. 04, pp. 437–490, 2012.

[WEI 89] WEINRICH H., *Grammaire textuelle du français*, Didier / Hatier, Paris, 1989.

[WER 75] WERLICH E., *Typologie der Texte*, Quelle & Meyer, Heidelberg, 1975.

[WIE 95] WIENER E., PEDERSEN J., WEIGEND A.S., "A neural net approach to topic spotting", *Fourth Annual Symposium on Document Analysis and Information Retrieval (SDAIR'95)*, pp. 317–332, 1995.

[WIE 05] WIEBE J., THERESA-WILSON CARDIE C., "Annotating expressions of opinions and emotions in language", *Language Resources and Evaluations*, vol. 39, nos. 2–3, pp. 165–210, 2005.

[WIL 85] WILKS Y., BRACHMAN R., SCHMOLZE J., "An overview of the KL-ONE knowledge representation system", *Cognitive Science*, vol. 9, no. 2, pp. 171–219, 1985.

[WIL 06] WILSON A., ARCHER D., RAYSON P. (eds), *Corpus Linguistics Around the World*, Rodopi, Amsterdam 2006.

[WIN 72] WINOGRAD T., *Understanding Natural Language*, Academic Press, New York, 1972.

[WIS 16] WISEMAN S., RUSH A.M., SHIEBER S.M., "Learning global features for coreference resolution", *Proceedings of the 2016 Conference of the North American Chapter of the Association for Computational Linguistics: Human Language Technologies*, Association for Computational Linguistics, San Diego, CA, pp. 994–1004, available at: http://www.aclweb.org/anthology/N16-1114, June 2016.

[WIT 53] WITTGENSTEIN L., *Investigations Philosophiques* (French translation in 1961, republication in 1986), Gallimard, Paris, 1953.

[WIT 05] WITTEN I.H., FRANK E., *Data Mining: Practical Machine Learning Tools and Techniques*, 2nd ed., Morgan Kauffman, San Francisco, CA, 2005.

[WON 09] WONG W., LIU W., BENNAMOUN M., "Acquiring semantic relations using the web for constructing lightweight ontologies", *13th Pacific-Asia Conference on Knowledge Discovery and Data Mining (PAKDD)*, pp. 266–277, 2009.

[WOO 75] WOODS W., "What's in a link: foundations for semantic networks", in BOBROW D., COLLINS A. (eds), *Representation and Understanding: Studies in Cognitive Science*, Academic Press, New York, 1975.

[WOO 95] WOOLDRIDGE M., JENNINGS N.R., "Intelligent agents: theory and practice", *Knowledge Engineering Review*, vol. 10, no. 2, pp. 115–152, available at: http://www.csc.liv.ac.uk/~mjw/pubs/, 1995.

[XIA 01] XIA F., PALMER M., "Converting dependency structures to phrase structures", *1st International Conference on Human-Language Technology Research*, pp. 61–65, 2001.

[YAN 00] YANGARBER R., GRISHMAN R., TAPANAINEN P. *et al.*, "Automatic acquisition of domain knowledge for information extraction", *Proceedings of the 18th International Conference on Computational Linguistics COLING00*, Universität des Saarlandes, Saarbrücken, Germany, 31 July–4 August 2000.

[YE 14] YE M., *Data Mining: Theories, Algorithms, and examples*, CRC Press, Boca Raton, 2014.

[ZAM 00] ZAMPARELL R., *Layers in the Determiner Phrase*, Garland Publishing, New York, 2000.

[ZEL 03] ZELENKO D., AONE C., RICHARDELLA A., "Kernel methods for relation extraction", *Journal for Machine Learning Research*, vol. 3, pp. 1083–1106, 2003.

[ZHA 05] ZHAO S., GRISHMAN R., "Extracting relations with integrated information using kernel methods", *Proceedings of the 43rd Annual Meeting of the Association for Computational Linguistics ACL'05*, pp. 419–426, 2005.

[ZIP 49] ZIPF GEORGE K., *Human Behavior and the Principle of Least Effort*, Addison Wesley, New York, London, 1949.

[ZOL 67] ZOLKOVSKIJ A., MEL'CUK I., "O semanticeskom sinteze [Sur la synthèse sémantique]", *Problemy kibernetiki*, vol. 19, pp. 177–238 (French translation: T.A. Informations, 1970, no. 2, pp. 1–85.), 1967.

[ZWE 95] ZWEIGENBAUM P., BACHIMONT B., BOUAUD J. *et al.*, "A Multilanguage architecture for building a normalized conceptual representation from medical language", *19th Annual SCAMC*, New Orleans, 1995.

[ZWE 98] ZWEIGENBAUM P., BOUAUD J., BACHIMONT B. *et al.*, "Évaluation d'une représentation conceptuelle normalisée de comptes rendus médicaux en langue naturelle", *11ième congrès Reconnaissances des Formes et Intelligence Artificielle*, Clermont Ferrand, January 1998.

Index

A, B, C

adverbials, 128
anaphora, 64, 85, 131, 135, 139, 140, 142, 143, 146, 154, 158, 159, 161, 165
ARIANE (MT system), 205
ASCII (American Standard Code for Information Interchange), 25
assumed, 76, 104, 116, 127, 133–136, 173, 246
Big Data (BD), 234
bona fide boundaries, 73
Brulex (lexical DB), 49
centering theory, 160
circumplex model, 251
clustering, 220, 221, 223, 225
cognitive models, 81, 180
coherence, 35, 65, 125, 137, 138, 148, 160–162, 174, 190, 235
cohesion, 125, 138, 139
combinatorial semantics, 75
comment, 127, 130, 131, 192, 248
conceptual graph, 54–57
connotation, 2, 24, 253
constant topic progression, 131, 133
cotext, 1, 128, 129, 240
CSTAR (MT project), 205

D, E, F

data warehouse, 235, 236
DEB (Dictionary Editor and Browser), 43
denotation, 2, 13, 22, 158
derived topic progression, 131, 132
discourse
  analysis, 123, 125, 128, 146, 148, 151, 153, 181
  interpretation, 154
DOLCE (Descriptive Ontology for Linguistic and Cognitive Engineering), 69
ellipses, 142, 143, 189, 191
empty words, 216, 229, 231
EUROTRA (MT system), 196
event extraction, 247
example-based translation, 206
extract load transform (ELT), 235
frames, 32, 57–60, 212

G, H, I

generative
  lexicon theory, 21
  semantics, 80
hadoop, 236, 237
homography, 10

homonymy, 10, 11, 192
illocutionary acts, 145
implicit, 120, 133–135, 248, 249, 254
information
  extraction, 146, 177, 234, 238, 239, 243, 247
  retrieval, 45, 57, 147, 148, 177, 189, 211, 213, 219, 220, 228, 229, 234, 238, 256, 257
inserted topic progression, 131, 133
interlanguage, 40, 41, 51
interpretive semantics, 6, 75, 76, 78–80, 84, 90
intertextuality, 128, 130

L, M, N

lambda calculus, 113–116, 118, 120, 184
Latent Semantic Analysis (LSA), 147, 228
lemmatization, 219, 221, 223, 231
lexical field, 6, 139
linear topic progression, 131, 132
linguistic software, 170, 177
LMF (standard lexical DB), 36
locutionary acts, 145
meaning-text theory, 42
METAL (MT system), 195
metaphor, 3, 4
metonymy, 5, 239
multi-agent architectures, 178
named entities, 48, 239, 247
narrative text, 91, 124, 131, 143

O, P, R

ontologies, 43, 49, 50, 63, 65, 66, 68, 69, 71, 73, 171, 243
opinion extraction, 249, 255
opposition, 10, 11, 13–15, 123, 124, 135, 150, 217
paronymy, 15
perlocutionary acts, 145
polysemy, 6, 10–12, 15, 46, 192
pragmatics, 76, 123, 146
presupposition, 79, 80, 134, 135, 158
propositional logic, 95, 96, 106
prototype semantics, 18, 20
RDF (Research Description Framework), 31, 32, 45, 67
rhetorical structure theory, 148

S, T, U

SALT (project), 35
scripts, 37, 60, 61, 147
semantic
  field, 6, 91, 139, 217
  networks, 43, 51–54, 57, 61, 63
  transfer, 203, 205
SGML (Standard Generalized Markup Language), 25
signature files, 220
speech act, 143–147
SUMO, 69, 71, 72
synecdoche, 5
synonymy, 12, 13, 47, 139, 152, 228
syntactic transfer, 202
Systran (MT system), 195
TAUM (MT system), 195
TEI (Text Encoding Initiative), 32
textual sequences, 143
tf-idf (Term Frequency-Inverse Document Frequency), 217, 218, 231
topic, 57, 91, 94, 130–133, 147, 163, 172, 249, 253, 255
troponymy, 16
UNL (Universal Networking Language), 61
utterance production, 124–126, 128, 134

V, W, X

Verbmobil (project), 208
WordNet, 38, 43, 45–47, 53, 69, 71, 254
XML (eXtensible Markup Language), 29

Other titles from

iSTE

in

Cognitive Science and Knowledge Management

2017

MAESSCHALCK Marc
Reflexive Governance for Research and Innovative Knowledge
(Responsible Research and Innovation Set – Volume 6)

2016

BOUVARD Patricia, SUZANNE Hervé
Collective Intelligence Development in Business

CLERC Maureen, BOUGRAIN Laurent, LOTTE Fabien
Brain–Computer Interfaces 1: Foundations and Methods
Brain–Computer Interfaces 2: Technology and Applications

FORT Karën
Collaborative Annotation for Reliable Natural Language Processing

GIANNI Robert
Responsibility and Freedom
(Responsible Research and Innovation Set – Volume 2)

GRUNWALD Armin
The Hermeneutic Side of Responsible Research and Innovation (Responsible Research and Innovation Set – Volume 5)

KURDI Mohamed Zakaria
Natural Language Processing and Computational Linguistics 1: Speech, Morphology and Syntax

LENOIR Virgil Cristian
Ethical Efficiency: Responsibility and Contingency (Responsible Research and Innovation Set – Volume 1)

MATTA Nada, ATIFI Hassan, DUCELLIER Guillaume
Daily Knowledge Valuation in Organizations

NOUVEL Damien, EHRMANN Maud, ROSSET Sophie
Named Entities for Computational Linguistics

PELLÉ Sophie, REBER Bernard
From Ethical Review to Responsible Research and Innovation (Responsible Research and Innovation Set - Volume 3)

REBER Bernard
Precautionary Principle, Pluralism and Deliberation (Responsible Research and Innovation Set – Volume 4)

SILBERZTEIN Max
Formalizing Natural Languages: The NooJ Approach

2015

LAFOURCADE Mathieu, JOUBERT Alain, LE BRUN Nathalie
Games with a Purpose (GWAPs)

SAAD Inès, ROSENTHAL-SABROUX Camille, GARGOURI Faïez
Information Systems for Knowledge Management

2014

DELPECH Estelle Maryline
Comparable Corpora and Computer-assisted Translation

FARINAS DEL CERRO Luis, INOUE Katsumi
Logical Modeling of Biological Systems

MACHADO Carolina, DAVIM J. Paulo
Transfer and Management of Knowledge

TORRES-MORENO Juan-Manuel
Automatic Text Summarization

2013

TURENNE Nicolas
Knowledge Needs and Information Extraction: Towards an Artificial Consciousness

ZARATÉ Pascale
Tools for Collaborative Decision-Making

2011

DAVID Amos
Competitive Intelligence and Decision Problems

LÉVY Pierre
The Semantic Sphere: Computation, Cognition and Information Economy

LIGOZAT Gérard
Qualitative Spatial and Temporal Reasoning

PELACHAUD Catherine
Emotion-oriented Systems

QUONIAM Luc
Competitive Intelligence 2.0: Organization, Innovation and Territory

2010

ALBALATE Amparo, MINKER Wolfgang
Semi-Supervised and Unsupervised Machine Learning: Novel Strategies

BROSSAUD Claire, REBER Bernard
Digital Cognitive Technologies

2009

BOUYSSOU Denis, DUBOIS Didier, PIRLOT Marc, PRADE Henri
Decision-making Process

MARCHAL Alain
From Speech Physiology to Linguistic Phonetics

PRALET Cédric, SCHIEX Thomas, VERFAILLIE Gérard
Sequential Decision-Making Problems / Representation and Solution

SZÜCS Andras, TAIT Alan, VIDAL Martine, BERNATH Ulrich
Distance and E-learning in Transition

2008

MARIANI Joseph
Spoken Language Processing

Printed and bound by CPI Group (UK) Ltd, Croydon, CR0 4YY

13/09/2026

14961021-0007